History and Speculative Fiction

John L. Hennessey
Editor

History and Speculative Fiction

palgrave
macmillan

Editor
John L. Hennessey
Lund University
Lund, Sweden

ISBN 978-3-031-42234-8 ISBN 978-3-031-42235-5 (eBook)
https://doi.org/10.1007/978-3-031-42235-5

Cover credit: Tithi Luadthong / Alamy Stock Photo

This Palgrave Macmillan imprint is published by the registered company Springer Nature Switzerland AG.
The registered company address is: Gewerbestrasse 11, 6330 Cham, Switzerland

Paper in this product is recyclable.

Acknowledgments: History and Speculative Fiction

I would like to thank the many contributors to this volume and those who provided critical feedback on the introductory chapter, especially Johan Höglund, Janne Lahti, Jonas Hansson, and Victoria Höög. I am also grateful to the anonymous reviewers for their helpful comments.

This book has been made open access thanks to generous contributions from the Per Anders and Maibritt Westrin Foundation and the Lund University Libraries Book Fund (Bokfonden).

Contents

Notes on Contributors

Ella Andrén is a PhD student in History at Linnaeus University. She has published articles on personal uses of history and popular culture in the web community debates on historical television series like *Downton Abbey*. She is also an editor of the website dagensbok.com, where she has published hundreds of literary reviews. Her research interests include historical consciousness, history education, and culture studies.

David Belden completed his doctorate in 1976 at Oxford University on the history and sociology of the Oxford Group, later known as Moral Re-Armament (MRA), the religious movement in which he had been raised and with which he had volunteered in Europe, the United States, India, and Ethiopia. He left MRA at age 23. He became friends with Gunlög Für, later as fellow dissenters from that movement. Now 73, he has worked as a writer, editor, carpenter, professor, and facilitator. His longstanding desire to combine personal healing with systemic change has led to his past ten years' work in restorative justice. In the 1980s, he had two science fiction novels published, *Children of Arable* and *To Warm the Earth*. He emigrated to the United States in 1981 to marry Debi Clifford, an American; they have a son, Rowan, and granddaughter Marlowe, aged 3.

Hans Hägerdal is Professor of History at the Department of Cultural Sciences, Linnaeus University. He completed his PhD thesis, *Väst om öst: Kinaforsknig och Kinasyn under 1800-och 1900-talen* [West on East: Sinology and Views of China During the Nineteenth and Twentieth Centuries], in 1996 (Lund University). He has since conducted research about colonial contact zones in Southeast Asia, indigenous history-making, and the history of slavery. He is currently engaged in a project about diplomatic history in Southeast Asia, 1700–1920 (projected for 2022–2026). He is the author, among other works, of *Lords of the Land, Lords of the Sea: Conflict and Adaptation in Early Colonial Timor, 1600–1800* (Brill, 2012).

Ashleigh Harris is Professor of English Literature at Uppsala University. She is the author of *Afropolitanism and the Novel: De-Realizing Africa* (Routledge, 2019) and is currently working on a monograph entitled *Literary Form Beyond the Book in Southern Africa*. Her recent research has been focused on literary forms circulating outside of the formal book and publishing industry in Sub-Saharan Africa. Recent research in this area includes "'The Diary of a Country in Crisis': Zimbabwean Censorship and Adaptive Cultural Forms," *Journal of Southern African Studies* (2021), and "Early Sesotho, isiXhosa and isiZulu Novels as World Literature" in *African Literature as World Literature*, edited by Alexander Fyfe and Madhu Krishnan (2022).

John L. Hennessey is Associate Professor of the History of Ideas and Sciences at Lund University. He has published on global colonial history and the history of science in journals including *Science in Context*, *History and Anthropology*, *French Colonial History*, *Settler Colonial Studies,* and *Japan Review*. His dissertation, *Rule by Association: Japan in the Global Trans-Imperial Culture, 1868–1912* (Linnaeus University, 2018), was shortlisted for the 2019 ICAS Dissertation Prize in Asian studies and received an honorable mention for the 2020 Walter Markov Prize in global history. His research interests include modern colonial history, history education, scientific racism, and propaganda.

Anna Höglund is Senior Lecturer in Comparative Literature at Linnaeus University, Sweden. Her research areas are horror fiction and fantastic fiction in literature and film, and her aim is to investigate how the fantastic in different cultures and societies concretizes relations with the surrounding world in forms of needs, attitudes, and norms. She has a particular interest in the functions of monsters like vampires, zombies, and werewolves, as instrumental interpretations of the world. Höglund teaches classes on topics like vampire fiction, children's and horror fiction, and studies of power in the *Game of Thrones* series.

Johan Höglund is Professor of English and a member of the Linnaeus University Centre for Concurrences in Colonial and Postcolonial Studies,

Sweden. He has published extensively on how popular culture narrates and interrogates colonialism, neocolonialism, and extractive capitalism. He is the author of *The American Imperial Gothic: Popular Culture, Empire, Violence* (Routledge, 2014) and the co-editor of several scholarly collections and special journal issues, including *Dark Scenes from Damaged Earth: Gothic and the Anthropocene* (UMinn Press, 2022), *Nordic Gothic* (Manchester UP, 2020), "Nordic Colonialisms" for *Scandinavian Studies* (2019), *B-Movie Gothic* (Edinburgh UP, 2018), and *Animal Horror Cinema: Genre, History and Criticism* (Palgrave Macmillan, 2015).

Karen Ordahl Kupperman is Silver Professor of History Emerita at New York University. Her latest book, *Pocahontas and the English Boys: Caught Between Cultures in Early Virginia*, was published by NYU Press in 2019, along with a new edition of Henry Spelman's *Relation of Virginia* from the original manuscript. She is the author of *The Jamestown Project* (Harvard, 2007) and *The Atlantic in World History* (Oxford, 2012). Among her many awards are the AHA Prize for the best book in Atlantic History (2001) and the AHA's Beveridge Prize for the best book in American history (1995). Her current book project is on life in North America before the arrival of Europeans.

Janne Lahti works as an Academy of Finland Research Fellow at the University of Helsinki. He has been awarded a Fulbright Fellowship, Huntington Library fellowships, and Kone Foundation research funding, among others. His research focuses on global and transnational histories of settler colonialism, borderlands, the American West, German colonialism, and Nordic colonialism. He has published numerous articles, edited special issues, and also authored or edited eight books, including *Finnish Settler Colonialism in North America* (with Rani-Henrik Andersson; Helsinki University Press, 2022), *Kolonialismi Suomen rajaseuduilla* (with Rinna Kullaa and Sami Lakomäki; Gaudeamus, 2022), *German and United States Colonialism in a Connected World* (Palgrave, 2021), *Cinematic Settlers* (with Rebecca Weaver-Hightower; Routledge, 2020), and *The American West and the World* (Routledge, 2019).

Kristín Loftsdóttir is Professor of Social Anthropology at the University of Iceland. Her research has focused on notions of exceptionalism, the racialization of mobility, racism, gender, crisis-talk, globalization, nationalism, migration, and postcolonial Europe. She has published extensively on these issues in respected international journals and in Iceland. Her publications include *We Are All African Here: Race, Mobilities and West Africans in Europe* (Berghahn, 2021), *Exceptionalism* (co-author, Routledge, 2021), *Crisis and Coloniality at Europe's Margins: Creating Exotic Iceland* (Routledge, 2019), and *Messy Europe: Crisis, Race and the Nation State in a Postcolonial World* (co-editor, Berghahn, 2018).

Piia K. Posti is Senior Lecturer in Comparative Literature at Linnaeus University. She has published on travel writing, travel and exoticism in children's literature, and popular fiction. Recently, she has co-edited the anthology *Speglingar av feelgood: Genre, etikett eller känsla?* [Reflections on Feelgood: Genre, Label or Feeling?] (2022) in which she also writes on the genre classification of feelgood literature and the feelgood Christmas novel. Her current research interests include race and the impact of post-colonial research on popular media adaptations and popular fiction.

Cecilia Trenter is Professor in History at Malmö University. She works within the research field of memory studies and public history, including heritage adaptions and remediation in fiction, for instance, epic films (Zack Snyder's *300*) and computer games (BioWare's *Dragon Ages* series). In addition, she is studying heritage sites with educational approaches, for instance, in the article "Collective Immersion by Affections: How Children Relate to Heritage Sites" (Trenter, Ludvigsson & Stolare, *Public History Review* 28, 2021). Trenter is co-editor, with Anna Höglund, of *The Enduring Fantastic: Essays on Imagination and Western Culture* (McFarland, 2021) and, with Björn Horgby, of *I Folkets namn. Högerpopulism på film* [In the Name of the People: Right-wing Populism on Film] (Carlssons förlag, 2021).

Martin van der Linden is a doctoral student in the study of religions at Linnaeus University, Växjö. His dissertation project studies Shinto and the Shintoesque on the Internet seeking to analyze and understand what forms of aesthetic, multimodal, and intermedial patterns and structures disseminate and circulate information about Shinto online and what forms of ideological notions such patterns may carry. He is part of the doctoral research training program in Global Humanities at Linnaeus University and is a member of the Linnaeus University Center for Intermedial and Multimodal Studies (IMS).

CHAPTER 1

Introduction to History and Speculative Fiction: Essays in Honor of Gunlög Fur

John L. Hennessey

> Anyone can identify what seems odd or false in the mental habits of an alien "somewhere." But if something is the very texture of any insider's thought, anywhere, it is the work of genius, not of ordinary men and women, to think that one's thought is wrong. That is why I suggest that it is important to complement... [the] strategy of making the strange familiar, with the opposite one of making the familiar strange. (Fields and Fields 2012, 223)

History and speculative fiction have different epistemological starting points; simply put, history is based on fact and speculative fiction, as manifested by both parts of its name, is not. Nevertheless, when at their best, the two perform similar work in "making the strange familiar" and "making the familiar strange" by taking their readers on journeys through space and time. Excellent history, like excellent speculative fiction, should cause us to reconsider crucial aspects of our society that we normally overlook or else help us to break free of such discursive constraints through the process of familiarizing ourselves with radically different forms of social organization, whether in the factual past or the fictional future (or past or present).

J. L. Hennessey (✉)
Lund University, Lund, Sweden
e-mail: john.hennessey@kultur.lu.se

J. L. Hennessey (ed.), *History and Speculative Fiction*,
https://doi.org/10.1007/978-3-031-42235-5_1

This applies especially to the subtle structures of power that organize our own societies, whether through notions of gender, race, coloniality, or others that we do not so readily imagine. These structures become most apparent in liminal spaces between cultures or in contacts between societies, a phenomenon that Gunlög Fur has explored through the concept of *concurrences*. As will be described in more detail below, *concurrences* describes separate, parallel worlds or cultures that operate according to different internal logics, but come into contact, generating complicated relations of agreement and/or competition (Brydon et al. 2017a).

How varied can the organization of human society be, and what are the common denominators between different cultures and lifeways? How do we define the "human" and imagine its relationship to what we accordingly define as "non-human"? Are the different societies imagined by the authors of speculative fiction viable in real life (whatever that means) or too "inhuman" to be plausible? To what extent are historians able to immerse themselves in and understand the world of past individuals who lived in vastly different social arrangements, and to what extent are they limited in this endeavor by their own cultural baggage?

These questions pose great challenges, but working through them also holds the promise of exposing unjust and discriminatory power structures within societies, promoting understanding between societies and maybe even preparing humanity to cope with crises, whether pandemics, climate change, or currently unimaginable future issues. In this, history and speculative fiction, which have hitherto seldom been considered together (at least from the history side), may be able to learn from each other and even become allies. Despite their different premises, is the knowledge generated by each somehow compatible or complementary? How might historians become better equipped to study the past through a consideration of fictional societies, and how might authors of speculative fiction write better works with more nuanced understandings of history? How might a more profound understanding of historical approaches help literary scholars in their work?

With contributions from a variety of disciplinary perspectives that consider diverse examples of speculative fiction and historical encounters, this volume provides a robust opening to a serious discussion of these questions. At a time that the discipline of history has been described as being in deep crisis even as historical claims are increasingly mobilized in the service of political battles and identity creation (Bessner 2023), a productive engagement with speculative fiction may provide one avenue for reinvigorating the discipline and creatively addressing future challenges. The authors of this volume hope in any case that this book will be a fitting tribute to 30 years of groundbreaking scholarship and conscientious teaching by our colleague, mentor, and friend Gunlög Fur, not least through new applications of the concept of *concurrences*.

Concurrences

> To see and name emergent patterns of globalization, and to look again at the histories that have brought the world to this place, requires experimental methodologies that proceed from fresh assumptions about modes of knowing and value. (Brydon et al. 2017a, 30)

The present volume shares the same overarching concern expressed here by Diana Brydon, Peter Forsgren, and Gunlög Fur in finding creative new methods to understand human societies. The method that these two literary scholars and one historian argue for in the anthology from which this quote is taken, *Concurrent Imaginaries, Postcolonial Worlds: Toward Revised Histories* (2017b), is Gunlög Fur's postcolonial concept/methodology of *concurrences*. One of the key aims of *History and Speculative Fiction*, both here and in the chapters that follow, is to demonstrate that concurrences and speculative fiction are especially productive in combination, in several different ways. Both center around the meeting and evaluation of different cultures, languages, and ways of life. As described in more detail below, science fiction has strong ties to historical colonialism and, like the postcolonial concept of concurrences, is unusually well-suited to critique its legacies. Moreover, history and speculative fiction can be understood as concurrent modes of exploring the human condition and shaping our view of possible futures. Perhaps most importantly, speculative fiction is an especially apt tool for helping us to understand the meaning and significance of concurrences and explore its potential, while concurrences as a critical historical method offers the possibility of enriching and overcoming the colonial tropes that still shape much science fiction.

But first it is necessary to better explain what *concurrences* is. As Brydon, Forsgren, and Fur themselves admit (2017b, 15), concurrences can be challenging to understand abstractly. The concept was developed to better explain the meeting of two different cultures, epistemologies, or value systems, a situation that frequently arose historically in the context of colonialism. Such meetings were particularly frequent and important in the early modern contact between Europeans and Native Americans, Fur's primary area of expertise. As Fur herself explains,

> "Concurrences"… refers to disparate spheres of existence and meaning that are interlinked but do not necessarily overlap and are not organized hierarchically—even though asymmetrical power-nexuses will influence these relations. The nature and evolution of these power-relations, however, are questions of historical study and context, not an organic or essential (or theoretically predictable, or even predestined) aspect of these relations. (2017, 54)

Concurrences seeks to understand this contact between different "spheres of existence and meaning" without privileging one or the other, but also without naïvely ignoring the presence of the very real unequal power relations that such meetings often involve. As an approach, it does not view these inequalities as inevitable or reflecting the essence of either sphere, however; to do so would risk perpetuating colonial tropes like that of the "dying race" (see, for example, Brantlinger 2003) or the superiority of certain forms of "civilization." "To think in terms of concurrences is to reject both binary models of opposition and absolute models of relativism," as Brydon, Forsgren, and Fur put it (2017b, 11). Concurrences aims therefore to complicate our view of the world by bringing in multiple perspectives, while not falling victim to either absolute relativism or conceiving of these meetings on an idealistic plane outside of real-life power differentials on the one hand, or oversimplified, stereotypical, or Manichean views of cultural difference on the other.

Concurrences is not about a meeting on "neutral ground," but often involves competing claims, or what Fur frequently refers to as "jurisdictions":

> It is not the multiplicity of histories per se that interests me but the way in which they become entangled, ensnared by their competing jurisdictions. *Concurrences* points to those zones of entanglement where simultaneous presence in time and space reveals not only separate claims on jurisdiction but also how people deal with difference and similarity, closeness and distance, in ways that belie simplistic categorizations and predetermined hierarchies. (Fur 2017, 46)

As such, the notion of *place* or spatiality is central to concurrences, both in terms of the real-world situatedness of such encounters and in the importance of the place from which the scholar is researching and writing. Like many postcolonial approaches, concurrences argues that researchers should pay especially close attention to their own specific baggage and remain humble to the fact that they will only be able to see an incomplete picture of the phenomena they are studying (Fur 2017, 40). This applies to all of the different epistemological positions, "worlds," or "cultures" involved in concurrent meetings; echoing Donna Haraway, Fur makes sure to point out that it is a mistake to romanticize "the other" or succumb to the temptation of "uncritically favouring subjugated or subaltern perspectives" (2017, 49).

Fur chose the term *concurrences* because it contained a richness of meaning, with different connotations that capture the different perspectives and approaches described above. Besides the most common present-day meaning of "simultaneous," *concurrence* can signify both agreement or (in its archaic English form or current Swedish form) competition, reflecting the different possible results of contact between different worlds.

> A term such as *concurrences*, then, contains in its bag of meanings both agreement and competition, entanglement and incompatibility as it slides uneasily across time ("archaic" noun-forms) and space (different languages). It signals contestations over interpretations and harbours different, diverging, and at times competing claims that will inflect studies of things such as home, travelling, subjectivity-identity, voice, and space. (Fur 2017, 40)

Like much speculative fiction, Fur's concept of concurrences suggests that understanding other cultures, lifeways, languages, and epistemologies that one comes into contact with is difficult, but possible (within certain limits set by one's own cultural baggage), and above all, important.

Many works of speculative fiction dramatize the meeting of mutually incomprehensible societies or worlds described by *concurrences*. In science fiction, difference is typically represented by alien species, as effectively demonstrated by Ella Andrén in her chapter on *Star Trek*. Even if Babel fish or universal translators are a frequent convenience in science fiction that allows authors to circumvent, instead of exploring, the difficulties of translation and epistemology involved in concurrent encounters, there are still countless examples in which these very difficulties form the crux of the story. This is a central theme in Charlie Jane Anders' novel *The City in the*

Middle of the Night (2019), explored in Karen Ordahl Kupperman's chapter, which draws insightful parallels between the regimented colonial society of the planet January, that depicted in Harry Martinson's *Aniara* (1956) and that of colonial Jamestown. *Children of Time* (2015) and *Children of Ruin* (2019) by Adrian Tchaikovsky, the *Binti* stories by Nnedi Okorafor (2015), and *A Memory Called Empire* (2019) and *A Desolation Called Peace* (2021) by Arkady Martine are only a few other recent examples that come to mind. Even *Star Trek: The Next Generation*, with its heavy reliance on universal translators, dramatized the difficulty of understanding a completely different way of communication (however implausible) in the episode *Darmok* (1991). These examples mostly have happy endings in which some form of mutual understanding is established, but there is also a tradition in science fiction that emphasizes the impossibility of understanding alien others. This is true of Stanisław Lem's *Solaris* (1961), Arthur C. Clarke's *Rendezvous with Rama* (1973; at least before the sequels were written), and Arkady and Boris Strugatsky's *Roadside Picnic* (1972) [2012]), all of which involve encounters with extremely powerful, but completely inscrutable aliens.

I would argue that one of the best dramatizations of *concurrences* in speculative fiction, and one that can help us to better understand and reflect on the term itself, is Ted Chiang's *Story of Your Life* (2002, originally published in 1998). This award-winning novella was the basis of the film *Arrival* (2016), but the original story has certain key differences. In a few dozen pages, the story richly connects profound reflections on linguistics, parenthood, the nature of time, and epistemology. But it is the contact between two previously isolated species or civilizations and their worldviews that is the most relevant for a discussion of concurrences. In the story, humanity is unexpectedly visited by an advanced alien race, which they call "heptopods." The heptopods remain in orbit around Earth but send down 112 "looking glasses"—a kind of two-way audiovisual communication device through which they come into contact with humanity. The story centers on the narrator, Louise Banks, a linguist, and Gary Donnelly, a physicist with whom she is paired in order to establish communication with and study the heptopods, under the command of the U.S. Military. The military officers and other representatives of the U.S. Federal Government represent a narrow-minded, binary approach to otherness. They are completely flummoxed by the intentions of the heptopods, whom they perceive first as a military threat and later, greedily, as a potential source of advanced technology. Frustrated by the government's lack of flexibility and creativity in this first human contact with an alien species, the researchers, who are driven primarily by a will to understand the heptopods, come to a far deeper, though still incomplete, understanding of these visitors.

As Louise slowly comes to comprehend the heptopods' significantly different language, one of Gary's physicist colleagues finally makes a breakthrough after weeks of being unable to communicate about physics concepts when the heptopods react with understanding to a description of Fermat's principle. The resulting physics discussion is (predictably) omitted in the Hollywood movie but is arguably essential to the plot of the story. In discussing Fermat's principle, which describes how light always "chooses" the fastest path between two points, even when traveling through different, refracting, mediums, a key difference between human and heptopod epistemology comes to light. Donnelly explains that laws of physics are typically expressed in causal terms, but that mathematically, it is just as correct to describe them in other terms, as variational principles: "The thing is, while the common formulation of physical laws is causal, a variational principle like Fermat's is purposive, almost teleological" (124). As stated even more clearly later,

> The physical universe was a language with a perfectly ambiguous grammar. Every physical event was an utterance that could be parsed in two entirely different ways, one causal and the other teleological, both valid, neither one disqualifiable no matter how much context was available. (133)

It dawns on Louise that the heptopods' language and physics reflect an entirely different epistemology and way of relating to time: humans have a "sequential mode of awareness" and heptopods a "simultaneous mode of awareness" (134).

As Louise increasingly masters their written language, she begins to think like the heptopods and suddenly has access to her own "memories" from the future. Unlike in the movie, however, Louise realizes that she cannot use her knowledge of future events to affect them:

> Freedom isn't an illusion; it's perfectly real in the context of sequential consciousness. Within the context of simultaneous consciousness, freedom is not meaningful, but neither is coercion; it's simply a different context, no more or less valid than the other... But you can't see both at the same time. Similarly, knowledge of the future was incompatible with free will. What

> made it possible for me to exercise freedom of choice also made it impossible for me to know the future. Conversely, now that I know the future, I would never act contrary to that future... (137)

In the story, this is not a "gift" from the heptopods, per se, but a result of Louise understanding their worldview, albeit one that she cannot completely share. A mutual exchange does play a role in the story, however. At the insistence of the researchers, the government decides to eschew attempts at trade and instead engage in mutual "gift-giving" with the heptopods. In the movie, the heptopods intentionally "gift" humans the ability to see into the future as a kind of quid pro quo arrangement (so that humanity will be able to help the heptopods in the future), but in the story, the gift giving is much freer, more whimsical, and of less perceived value to the U.S. government. In the end, the heptopods leave as mysteriously as they arrived; Louise has gained new perspectives on the universe but still does not fully understand the heptopods or their motives.

In many ways, Chiang's story demonstrates the concept of *concurrences.* The heptopods consistently resist human governments' attempts to place them into predefined categories or understand them according to human cultural logic; it is only through curiosity, openness, and a cognizance of their own subject positions and cultural embodiments that researchers like Louise are able to come to a greater understanding of them. The gift-giving paradigm that the heptopods positively respond to is also emphasized in the theorization of concurrences. Based on calls from indigenous scholars, gift-giving is highlighted as "a proper stance for academic intercourse"—the free exchange of stories and knowledge (Fur 2017, 41). Most interestingly, Chiang's story, like much of his work, can be characterized as "hard science fiction," engaging in an informed way with physics, linguistics, and mathematics. And yet, *Story of Your Life* still embraces the possibility of multiple but equally "true" worldviews, even when it comes to fundamental scientific principles. The story therefore not only illustrates *concurrences* in a particularly nuanced and striking way but also demonstrates how multiple epistemologies or worldviews need not be merely the fantasies of "soft" subjects within the humanities and social sciences but that the social construction of even the "reality" of the universe may be mathematically plausible. Explicitly or implicitly, the rest of the chapters of this volume explore the productive synergies between history and speculative fiction through the lens of concurrences.

Distinguishing Between "History" and "Speculative Fiction"

Concurrences, then, is arguably a productive way of conceiving of the relationship between history and speculative fiction, but implies that they are separate logics or fields that can intersect in complex ways. Such a sharp distinction, however, is not uncontroversial. The difference between history and fiction has been the subject of debate for centuries, although the attempt to make history a more "scientific" discipline in the nineteenth century is generally seen as a turning point, with the creation of a sharper boundary between the two (see, for example, Burke 2012). More recently, the debate flared up and became the subject of countless articles and books in the final decades of the twentieth century, with Hayden White as a major figure of controversy. As David Carr has pointed out, both the positivist defenders of historical "objectivity" and many critics, like White, who argue that history is inherently more literary than these positivists would like to admit, share the same assumption that "creative" or "literary" elements in historical studies are suspect and that fiction is analogous to falsification or deception (2004). In fact, as Carr astutely argues, novelists are hardly deceptive, as it is clear from the context in which their works are read that they are not intended to be taken as "true." Nor does the use of literary elements automatically invalidate historical research; the distinction is rather one of the intentions. Both history and fiction can use similar techniques, but history is characterized by its production of "assertions, theories, predictions, and in some cases narratives, about how the world really is, or will be, or was," while fiction is not—a distinction which, according to Carr, nearly all readers understand (2004, 255).

Despite the massive literature on the relationship between history and literature, or fiction, in general, *speculative fiction* or related categorizations such as science fiction and fantasy have received virtually no consideration in the context of historical methodology or epistemology. Since speculative fiction cannot be mistaken for a factual account of the world, it has been overlooked in the aforementioned discussions of history and literature (for a rare exception, see Liedl 2015). Nevertheless, as Carr contends, such discussions have largely missed the point of how both history and literature can shed light on the human experience in different, often complementary, ways. Speculative fiction's unrealistic nature can actually make it particularly useful for understanding the nature of historical truth and why scholars believe in certain facts. As Brian Attebery has argued for the fantasy genre, "Because fantasy has those irreducible elements of the impossible, the unreal, the extremely extraordinary, it helps us understand better what's the possible, what's the real, what's the true" (2022). The essays in this volume provide clear examples of the productive synergies between academic history and speculative fiction that can enhance historians' research and teaching.

Speculative fiction can be broadly defined as literature of the fantastic, using clearly unrealistic elements to explore hypothetical scenarios or bring aspects of the reader's world into sharp relief. It overlaps to a great degree with science fiction but can also be considered a broader, umbrella category that includes fantasy literature, which does not have the same focus on technology that typically characterizes science fiction. Science is often, but not always, the main focus of speculative fiction, and many creative stories take place in low-technology societies in the distant past or future that are at least as thought-provoking as literature involving high technology. In her exploration of postcolonialism and science fiction, Jessica Langer argues against the term "speculative fiction" in favor of "science fiction," which she feels better highlights the dark sides of scientific "progress" and its "conflict" with indigenous epistemologies that are often criticized in postcolonial scholarship (2011, 9). This is a valid point in the context of Langer's book, but even though much of the present anthology discusses colonialism, it is not limited to this topic, and I contend that "speculative fiction" is more useful when discussing synergies between this kind of literature and history-writing in general.

The emphasis on literature that is *speculative* highlights the intellectual, contemplative dimension of the best of this literature. In my view, "science fiction" is too-closely associated in everyday speech with space opera. I find it difficult to categorize works like *Star Wars* as "speculative" or engaging with important questions of how society is organized. For such "science fiction," spaceships, lasers, robots, and other high technology are mostly exotic scenery that could easily be swapped for sailing ships, castles, and horses. In speculative fiction, however, the fantastic elements form a crucial part of the plot and its *raison d'être*, making it perhaps a more serious (though not always less fun) type of fiction. For these reasons, this book will use both "speculative fiction" and "science fiction," but the former is preferred when discussing this kind of literature and its relationship to history in a more general way. In addition, the diverse chapters that follow are not limited to print literature, but explore different media used to convey creative speculation.

Reflecting the close affinity between these genres or modes of writing, two literary critics' explanations of what science fiction is apply equally well to speculative fiction. David Seed describes science fiction as "an embodied thought experiment whereby aspects of our familiar reality are transformed or suspended" (2011, 2). Darko Suvin has similarly characterized science fiction as "literature of cognitive estrangement," in which rigorous coherence in world-building according to fantastic premises is paramount (Suvin 1979, quoted in Rieder 2011, 62). This last definition of science fiction is particularly useful for the present discussion of science- or speculative fiction's connections to history, for cannot history also be described as a "literature of cognitive estrangement"? The great challenge for historians is becoming so immersed in the language and culture of the "foreign country" of the past as to be able to understand its obscure references and oblique jokes. Failure to properly do so could have disastrous consequences, with historians completely misunderstanding and misconstruing key texts, events, and processes.

I often try to explain to my students that the people of the past were not stupider than we are, even if they were ignorant of later developments and even though their worldviews can seem laughably wrong in our eyes. The deeply engrained narrative of explosive technological progress on which our modern identity rests tends to obscure the many things that the people of the past knew but that have now largely been lost. Myriad philosophical and religious ideas, once-canonical texts, social norms, and even basic knowledge of agriculture, nature, or the uses of various tools that were common knowledge in specific times and places have now fallen into obscurity.

Though this insight can be gained through the detailed study of a historical period, it is quickly and usefully dramatized in a great many time travel stories within science fiction. The time traveler, wary of being uncovered as an imposter and burned as a witch, or worse, must become extremely well-versed in the local language and culture, almost like a spy. This is particularly skillfully executed in the time-travel novels and short stories of Connie Willis, in whose universe time travel is the domain of academic historians conducting fieldwork, since the inability to alter the timeline has made it unprofitable and therefore uninteresting to commercial actors. In Willis' different stories, the theme of past alterity is depicted in varying registers, whether in the somber account of the struggles of a time traveler stuck in the Bubonic Plague in *Doomsday Book* (1992) or the comedy of errors that ensues from the protagonist's inadequate preparation for his mission to Victorian England in *To Say Nothing of the Dog* (1997). Infiltrating the society of the past in time travel narratives like Willis' not only forms an exciting narrative but also exposes the richness of the past and all that has been lost in a particularly vivid way. Even later historians' (or archaeologists') errors of interpretation are the subject of works like *A Canticle for Leibowitz* (Miller 1959) or *Motel of the Mysteries* (Macaulay 1979). In the former, a twentieth-century mechanic's shopping list becomes venerated centuries later as a holy relic. In the latter, a lampoon of the discovery of Tutankhamun's tomb, a future archaeologist completely misunderstands the purpose of the everyday objects he excavates from a twentieth-century motel room.

The confusing mix of familiarity and radical alterity in the past in many ways mirrors the construction of fantastical worlds in speculative fiction. In both historical sources and speculative fiction, the present-day reader can be lured into a sense of security by familiar cultural or material elements that both their own society and the society they are reading about hold in common before being jarred by an unexpected difference that reveals the similarities to be mostly superficial. Insight into the complexity and distinct internal logic of other societies and the ability to convey some of this through a richness of detail in world-excavation or world-building are what mark both high-quality history and speculative fiction. History aspires to the analysis and interpretation of the past made possible by a deep knowledge of its innumerable contextual minutiae, whereas speculative fiction attempts the creation of an imaginary world that involves fantastic elements, but with a consistency and complexity that makes it seem plausible. Both involve a large degree of "cognitive estrangement" from their author's internalized assumptions about how the world and society are.

After the postmodern turn, most historians have been increasingly wary of their ability to set aside their own cultural biases when evaluating source material and writing history. The historian's attempt at "cognitive estrangement" will always be imperfect, limited by their own culturally- and linguistically-determined cognition. Similarly, virtually all works of speculative fiction can be criticized for logical or internal inconsistencies, or the seeming implausibility of the world that they create, distracting from their intended message. But as incomplete or imperfect as they inevitably are, history and speculative fiction still both offer unique possibilities to question the seeming inevitability of aspects of our current society, our current world.

Counterfactual Speculation

There is little research into the possible creative synergies between history and speculative fiction in general, but a great deal has been written about more specific types of speculative fiction or from other perspectives that shed light on this topic in useful ways. There are naturally many literary histories *of* science or speculative fiction, which often provide useful insights into the origins of certain conventions that have shaped these overlapping modes (see, for example, Seed 2011; Luckhurst 2018). This is particularly true of colonialism, whose special relationship to science fiction requires its own section later on.

More directly, although it only represents one subcategory of speculative fiction, a body of literature has arisen around the study of counterfactual history. This has become an increasingly popular and influential literary subgenre and one with clear implications for the study of history (see, for example, Rosenfeld 2005; Evans 2013). In one of the most important recent studies of the counterfactual, Catherine Gallagher argues that although counterfactual history has existed for centuries, it has become widespread and a significant political tool only comparatively recently. Starting in the 1970s, counterfactual methodologies became the object of serious discussion in both the historical and legal professions, the latter to address issues of restitution for historical crimes. In literature, Gallagher contends that counterfactual history went from being a science fiction subgenre to a mainstream literary mode in the first decade of the twenty-first century, in part thanks to the popularity of simulated historical battles in the gaming world whose outcome was open to change (2018, 1). Indeed, as this book demonstrates, especially Piia Posti's and Cecilia Trenter's chapters on romance fiction, this development has not been limited to realistic historical fiction, but counterfactual, fantastic, or speculative elements that were previously limited to science fiction have increasingly been used within other genres, further transgressing the already blurry boundaries of speculative fiction.

Gallagher's timeline demonstrates that interest in counterfactual speculation grew concurrently in different fields, including academic history, law, and literature. Historians have a long history of skepticism towards counterfactual speculation, but as Gallagher notes, many have come to see this as a possible supplementary tool for the profession. The counterfactual is a useful tool for considering issues such as "the role of human agency and responsibility in history, the possibilities of historical justice and repair, and the coherence of identity—of individuals, nations, and peoples—through time" (2018, 4). Just as this anthology argues for speculative fiction in general, counterfactuals can provide a useful means for reflecting on some of the core issues that make history meaningful.

While counterfactual history is of obvious relevance to the historical profession, it is only one of many types of speculative fiction. As I discuss in my chapter on what I label *counterphysical* fiction, the dominant form of merely changing the outcome of a battle or other historical turning point does not go very far in challenging our established ways of thinking about the world or how society could be. This anthology seeks to move beyond this narrow focus of much existing research and explore how other fantastic or speculative elements can enrich the writing of history and vice versa.

Colonialism and Postcolonialism

It should be apparent to anyone familiar with the pervasive science fiction tropes of galactic empires, (alien) race warfare, and the settlement of other worlds that science fiction has a close relationship to and is often directly inspired by real-life colonialism. This has been the subject of several major studies that, while not having the same focus on history-writing as this volume, discuss many of the same issues and theories explored here. Perhaps the most important study of the relationship between science fiction and colonial ideology is John Rieder's *Colonialism and the Emergence of Science Fiction* (2008). Rieder, like a majority of literary scholars, argues that the height of European colonial expansionism in the late nineteenth century was also, not coincidentally, the formative period for the most familiar aspects of modern science fiction (2). As a result, early science fiction is deeply infused with colonial themes and ideology, characteristics which have tended to persist over time. As Janne Lahti's chapter in this volume demonstrates, even recent science fiction blockbuster films in many ways reflect and perpetuate settler colonial ideology.

Rieder identifies several "powerful ideological fantasies" that characterize both colonialism and much science fiction. The first of these is the "discoverer's fantasy" of the *terra nullius* that is actually inhabited by indigenous peoples. Rieder defines a "missionary fantasy" as the attitude that "Although we know that our arrival disrupts and destroys the traditional way of life here, we believe that it fulfills the deep needs and desires of all right-thinking natives." The "anthropologist's fantasy" temporally displaces contemporaneous indigenous peoples by considering them as living in the past, "in fact, to be our own past" (31–32). Finally, there is the colonial fantasy that limited natural resources are actually unlimited, highlighting the close relationship between colonialism, capitalism, and environmental destruction (37). While not discussed explicitly by Rieder, his very use of the term "fantasy" points to yet another telling link between colonial history and speculative fiction: colonial ventures were in many ways based on speculation (in both senses of the term) about the profitability or usefulness of foreign lands, speculation that was very often inflated by a lack of reliable information and greed-induced delusions (Varnava 2015). In this way, a great deal of real-life colonial history was in fact based on fantasy and speculation more than reality, even though the terrible consequences of colonial expansionism were very real.

Rieder's ideological fantasies of colonialism and science fiction reflect many of the main colonial ideologies exposed by leading postcolonial theorists. Mary Louise Pratt's classic *Imperial Eyes* (1992), for example, presents a detailed analysis of the "discoverer's fantasy," which she describes in terms of the titular "imperial eyes" that see what they want to see and envision a concrete domination of the colonized landscape. The "anthropologist's fantasy," as the core of the colonial worldview, is discussed by a great deal of postcolonial scholarship, but perhaps most notably in Dipesh Chakrabarty's *Provincializing Europe* (2000), with its insightful discussions of the use of historical time by colonial ideology. Chakrabarty famously argues that European colonial ideology "consigned Indians, Africans, and other 'rude' nations to an imaginary waiting room of history" (8) while attacking the strange, atemporal "universalism" claimed by Europeans.

For Rieder, the centrality of time to colonial ideology is strongly related to its prevalence and importance as a motif in science fiction. This is particularly true of the common time travel motif. For those who accepted colonial notions of "civilization" and progress, travel in space was often understood as a kind of time travel (76). Early European travelers to Japan, for example, saw in Japanese society a mirror of Europe's Middle Ages. Those who visited a variety of so-called "primitive" cultures around the world frequently described them as "Stone Age people." Indeed, the idea that we can learn about, or from, "our primitive ancestors" by studying present-day human groups who have been isolated from globalized modern culture is still extremely prevalent today. The leap from spatial to time travel in science fiction, therefore, was not so great in the late nineteenth century and arguably facilitated by colonial ideology.

Since colonized "Others" were often considered to be not only culturally different but the colonial explorer's *own past*, both anthropology and science fiction often investigate "to what extent the limitations and weaknesses of contemporary humankind are effects of social organization rather than qualities intrinsic to the species" (Rieder 2008, 77). In my view, this is exactly the kind of difficult question that both history and speculative fiction should be jointly contributing to answering. Both types of writing remain unavoidably bound to their authors' preexisting worldview but can to some extent break out of these constraints, albeit in different ways: history by exploring the artifacts of different cultures and modes of social organization that actually existed in the past, and speculative fiction by using or constructing a rationally operating, consistent, but unreal, world in which to test such ideas. Despite its colonial legacies, then, (and, perhaps, despite history's Eurocentric and nationalist legacies) speculative fiction has the potential to productively challenge existing hegemonic ways of seeing the world in ways that can help to address current and future problems.

Indeed, while emphasizing science fiction's colonial origins, Rieder points out that from its very beginnings in the late nineteenth century, it was used both to reinforce and to question, critique, and destabilize colonial ideology (10). Nineteenth-century protagonists of science fiction stories were often the weaker party in a cross-cultural/cross-temporal colonial encounter, as travelers to the distant future were awed by its awesome technology in much the same way as colonial subjects visiting the metropole were supposed to be. As in many other contemporaneous empires, Japanese colonial authorities, for example, arranged "sightseeing tours" of Japanese cities for leaders of anti-colonial resistance in Japan's empire in an attempt to overwhelm and intimidate them into submission (Matsuda 2003, 48–49; Hennessey 2018, 228). Even more dramatically, the recurrent trope of the invasion of Earth by technologically superior aliens closely mirrored actual (and often contemporaneous) colonial conquest but placed familiar, "modern," or "civilized" characters in the position of the victim with whom the reader was intended to sympathize.

Catastrophes are indeed often central to speculative fiction, which frequently makes use of the related mode of dystopia. Kristín Loftsdóttir's chapter astutely discusses different ways that such imaginative works are linked to real-world "crisis-talk," intervening in current political debates.

Discussing the motif of catastrophe in science fiction that undermines technological and civilizational optimism, Rieder importantly argues that

> such logical or emotional inversion of the fantasies of appropriation is not just an imaginary effect. Environmental devastation, species extinction, enslavement, plague, and genocide following in the wake of invasion by an alien civilization with vastly superior technology—all of these are not merely nightmares morbidly fixed upon by science fiction writers and readers, but are rather the bare historical record of what happened to non-European people and lands after being "discovered" by Europeans and integrated into Europe's economic and political arrangements from the fifteenth century to the present. (124)

In this way, the "fantastic" or "speculative" elements of literature involving futuristic alien invasions or similar plots are actually startlingly real. This is in fact the very theme of the research project "Surviving the Unthinkable: Ecological Destruction and Indigenous Survivance in North America and the Nordic Countries, 1600–2022," initiated by Gunlög Fur shortly before this book went to press. The project will study whether humanity as a whole can learn from the experience of resilience after devastation experienced by a great many indigenous peoples around the world to be able to better cope with the devastation wreaked by climate change (Olsson 2022).

Based on Rieder's observation, much early science fiction could be said to reflect colonialists' fears of falling victim to their own methods or perhaps even, in some cases, their guilty conscience. Indeed, the late nineteenth and early twentieth centuries were characterized by an obsession with alien invasion and other forms of race warfare not only in science fiction but in non-fictional and even "scientific" works warning of a decline in white manliness or virility as a result of "over-civilization" (the Eloi in H. G. Wells' 1895 *The Time Machine* come to mind) or even outright "race suicide" in the face of an invasion by non-white masses (Bederman 1995; Painter 2010). This particular discourse has in fact recently been reinvigorated by the anti-immigration extreme-right in many European and European-settler countries, who use the language of "invasion" or "replacement" and, explicitly or not, fear that non-whites will in some way repeat the colonial crimes historically committed by Europeans (Bracke and Aguilar 2020). Adopting the position of the colonized victim in science fiction, then, does not necessarily lead to greater empathy with the historical victims of colonialism, but such fantasies can actually strengthen an "eat or be eaten" sense of being threatened by the Other and provide a more socially acceptable setting in which to explore such fears.

Besides the colonial origins of many classic science fiction tropes, the genre is also frequently criticized for its strong links to Eurocentrism and normative "whiteness" (see, for example, Carrington 2016). Ashleigh Harris' chapter explores how globally circulating views of science and science fiction produced in Europe and America could reinforce notions of whiteness in apartheid South Africa. Fortunately, the last several years have witnessed an upsurge of interest in postcolonial speculative fiction that uses the mode's various tools to creatively undermine persistent colonial ideology. In the afterword to a pioneering collection from 2004, *So Long Been Dreaming*, Uppinder Mehan argues that speculative fiction is a necessary complement to critical history in order to complete the work of decolonization:

> postcolonial writing has for the most part been intensely focused on examining contemporary reality as a legacy of a crippling colonial past but rarely has it pondered that strange land of the future. Visions of the future imagine how life might be otherwise. If we do not imagine our futures, postcolonial peoples risk being condemned to be spoken about and for again. (270)

This is likewise the focus of the increasingly salient literary movement Afrofuturism, which seeks to ensure that Black people have a prominent place in imagined futures that have long been predominantly white and reflect a homogenized Western culture (Carrington 2016; Lavender 2019). Such postcolonial speculative fiction contests persistent colonial tropes of the "inevitable" extinction or assimilation of non-dominant languages, cultures, and peoples.

In her 2011 book *Postcolonialism and Science Fiction*, Jessica Langer argues that science fiction need not necessarily be colonial, in spite of its origins, but in fact has characteristics that can be especially useful for overcoming the corrosive legacies of colonialism in postcolonial societies. One such characteristic is the capacity of science fiction to explore and dramatize otherness in particularly striking ways, whether through alien encounters or, as Rieder notes, cyborgs (Langer 2011, 85; Rieder 2008, 111). "In science fiction," Langer points out, "otherness is often conceptualized corporeally, as a physical difference that either signposts or causes an essential difference, in a constant echo of zero-world [real world] racialization" (82). As discussed above, tropes of alien invasion can, in their simplest forms, simply be a thin veneer for racist fears of immigration and "replacement," but they can also problematize real-world stereotypes through their critical examination of what the "human" consists of in contrast to actual aliens. Langer sees postcolonial science fiction's subversive potential to lie in a productive use of the hybridity theorized by postcolonial theorist Homi Bhabha:

> Rather than shying away from these colonial tropes [of the Stranger and the Strange Land]... postcolonial science fiction hybridizes them, parodies them and/or mimics them against the grain in a play of Bhabhaian masquerade... Their very power, their situation at the centre of the colonial imagination as simultaneous desire and nightmare, is turned back in on itself. (4)

The violence of the colonial encounter cannot be undone, but science fiction is one way to turn its own tropes against it and expose its injustices.

Langer also makes the important point that the study of science fiction needs to move away from the dominance of Euro-American and English-language works by highlighting literature from other languages and cultures (11). Langer does so in her book by discussing Japanese science fiction, much of which has never been translated and therefore has received little attention in English-language scholarship. As she points out, even postcolonial studies' predominant focus on the former British and French Empires, particularly India, "fails utterly to take into account the diversity of postcolonial experiences" (11). While this volume also does not completely overcome this bias, Martin van der Linden's, Anna Höglund's, and Cecilia Trenter's chapters treat Japanese, Korean, and Nordic speculative fiction, respectively, and, importantly, Hans Hägerdal's chapter takes up "the inclusion of Europeans in legendary and even fantastic contexts" by historical Southeast Asians, reversing the colonial gaze. That chapter in particular can hopefully serve as inspiration for future studies of speculative fiction from non-Western perspectives.

Eco-criticism

As one of the defining issues of our time, climate change has naturally been the subject of a great deal of recent speculative fiction and related literary scholarship. Speculative fiction was something of a forerunner in this regard, and Johan Höglund's chapter in this volume argues that

Ursula LeGuin's 1971 novel *The Lathe of Heaven* presciently registers the violence done to ecology by colonial/capitalist society. Science fiction has been especially well-suited to exploring a future climate calamity, not only because of its focus on the future but because it has a long tradition of focusing on catastrophes and post-apocalyptic worlds. As Johan Höglund's chapter shows especially clearly, climate change is inextricably linked to both colonialism and capitalism (themselves closely intertwined), a connection that much critical speculative fiction has dramatized in particularly striking ways. Climate change or ecological devastation are therefore not separate phenomena but ones that are closely related to other forms of colonial destruction. The obliteration of people, ecosystems, lifeways, and epistemologies go hand in hand in the totalizing, chauvinist logic of colonialism.

As we have already seen, Rieder argues that colonial "history haunts science fiction's visions of catastrophe," which often works through the actual destruction of cultures and peoples in a futuristic, exotic setting (2008, 124). This not only reflects subconscious processes but often involves a deliberate identification between colonial and speculative genocide and other crimes. As Rieder points out, H. G. Wells quite explicitly draws parallels between the Martian invasion and the Tasmanian genocide in *The War of the Worlds*, for example (1898; Rieder 2008, 132). With catastrophe being as strong a motif in classic science fiction as technological optimism, Rieder argues that

> visions of catastrophe appear in large part to be the symmetrical opposites of colonial ideology's fantasies of appropriation, so much so that the lexicon of science-fictional catastrophes might be considered profitably as the obverse of the celebratory narratives of exploration and discovery, the progress of civilization, the advance of science, and the unfolding of racial destiny that formed the Official Story of colonialism. (123–124)

Science fiction, then, from its nineteenth-century beginnings was at least as frequently characterized by technological skepticism as enthusiasm, reflecting the major reconsideration of technology driven by present-day climate change.

Though science fiction is typically associated with shiny spaceships and complex machinery, a great many modern examples of eco-critical speculative fiction use more of a natural idiom. Amal El-Mohtar and Max Gladstone's 2019 novel *This is How You Lose the Time War* takes place amidst a temporal war across history between two sides with different visions of the future that they fiercely defend. The two sides or futures can be seen as metaphors or perhaps actual embodiments of the nature-technology (or perhaps, more fundamentally, nature-culture) divide, with one being characterized by a mechanical/cyborg/networked intelligence style while the other, Garden, is characterized by natural imagery (though equally ferocious as its opponent and able to manipulate time and genetics in staggeringly advanced ways). The latter, with its depiction of natural elements as both immensely powerful and open to a different kind of high technology than the nuts-and-bolts kind most associated with science fiction, has become an increasingly common mode as genetics has taken a more prominent place at the forefront of humanity's scientific imagination. Monsters, instead of invading from other worlds, increasingly are the result of twisted genetic experiments or else come to symbolize the reaction of a personified Nature against human overexploitation. Two of the essays in this collection, by Anna Höglund and Martin van der Linden, take up these themes, both coincidentally involving boars as symbolic, destructive forces in Asian cinema.

Speculative fiction has played a crucial role in helping contemporary society imagine the potentially catastrophic effects of climate change in a near future. As the depth and seriousness of the situation becomes increasingly well-recognized, however, many commentators have questioned whether doomsday scenarios do more harm than good by sapping people of the hope and optimism that they require to effectively tackle the problem. In what could amount to a paradigm shift in the subgenre, Kim Stanley Robinson has attempted to restore some sense of optimism with his 2020 novel *The Ministry of the Future*. This novel depicts a potentially realistic (as in, not relying on "miracle" technologies) future in which humanity manages to bring down carbon emissions fast enough to mitigate the worst effects of climate change, even while suffering several massive catastrophes. It is unusual for speculative fiction to be so detailed and practically oriented, as a kind of a potential road map for overcoming climate change, blurring the boundaries between fiction and non-fiction, but the success of Robinson's novel may herald more such works in the near future. Though using more fantastic elements, this anthology's final contribution, an original short story by David Belden, similarly explores how humanity can productively work towards a more hopeful future.

"Free Your Mind"?

The imperative to "free your mind," as expressed by Morpheus in *The Matrix* (1999), is a common one in both critical academic history and speculative fiction. This introduction has argued that the two genres, fields, or modes of writing have much in common and stand to mutually profit through a deeper and more deliberate dialogue. Both offer the potential, using different strategies, to reach alternative understandings of human society through creatively engaging with the past and possible futures. How have humans lived before, and how might they live in the future? But the obvious related question, "How *should* humans live?" is a harder one, and one that academic history has often shied away from, in spite of the normative nature of much critical theory. In much the same way as envisioned by discourse theorists, examining the past and imagining the future can help us to break free of the epistemological, cultural, and cognitive limitations to which we all inevitably belong, if not completely so.

But what exactly is freedom? We live in an age often characterized as "neo-liberal," literally of "new freedom," and the well-off among us have unprecedented freedom to travel, consume, and mold our own identities. And yet, the unsustainability of our lifestyles has increasingly called into question the desirability of this form of freedom. Might it be that a knowledge of alternatives opens up for a new kind of freedom or desirable way of life? Just as Louise in *Story of Your Life* discovers that "knowledge of the future was incompatible with free will," could such alternative ways of thinking involve new responsibilities or burdens? Acknowledging climate change and the evils of colonialism is indeed a burden, but perhaps an awareness of alternate lifeways, whether from human history or speculation, can provide us with some degree of agency in shaping a desirable future as our present form of social organization becomes increasingly untenable.

The remainder of this book is divided into four sections. The first, "Colonialism, Oppression and Concurrences," directly addresses colonial and other forms of injustice as depicted or engaged with in speculative fiction, drawing heavily on the concept of *concurrences*. The second, "Alternative Histories, Alternative Realities," looks at the particularly direct engagement between history and speculative fiction through the counterfactual, or counterphysical. The third, "Defining and Defying the Boundaries of Cultures and the Human," investigates eco-critical speculative fiction with historical themes and the concurrent, uneasy relationship between the realms of nature and human culture, along with works that investigate the very nature and limits of what humanity is. The final section, "History, Speculative Fiction and Real-World Social Change," investigates how history and speculative fiction can be tools of activism in the present-day, real world. The final chapter in that section, and the book, is an original short story by science fiction author David Belden, highlighting how both types of writing can productively engage with one another.

We would like to express our deep gratitude to Gunlög Fur for her inspiration, generosity, and friendship over many years, and hope she will enjoy this book.

References

Anders, Charlie Jane. 2019. *The City in the Middle of the Night*. New York: Tor Books.

Attebery, Brian. 2022. The Best Books on Fantasy's Many Uses. Interview by Romas Viesulas. *Five Books*, November 25. https://fivebooks.com/best-books/fantasys-many-uses-brian-attebery/

Bederman, Gail. 1995. *Manliness and Civilization: A Cultural History of Gender and Race in the United States, 1880–1917*. Chicago: University of Chicago Press.

Bessner, David. 2023. The Dangerous Decline of the Historical Profession. *New York Times*, January 14.

Bracke, Sarah, and Luis Manuel Hernández Aguilar. 2020. "They Love Death as We Love Life": The "Muslim Question" and the Biopolitics of Replacement. *British Journal of Sociology* 71 (4): 680–701.

Brantlinger, Patrick. 2003. *Dark Vanishings: Discourse on the Extinction of Primitive Races, 1800-1930*. Ithaca: Cornell University Press.

Brydon, Diana, Peter Forsgren, and Gunlög Fur, eds. 2017a. *Concurrent Imaginaries, Postcolonial Worlds: Toward Revised Histories*. Leiden: Brill.

———. 2017b. What Reading for Concurrences Offers Postcolonial Studies. In *Concurrent Imaginaries, Postcolonial Worlds: Toward Revised Histories*, ed. Diana Brydon, Peter Forsgren, and Gunlög Fur, 33–58. Leiden: Brill.

Burke, Peter. 2012. History, Myth, and Fiction: Doubts and Debates. In *The Oxford History of Historical Writing*, ed. Andrew Feldherr et al., 261–281. Oxford: Oxford University Press.

Carr, David. 2004. History, Fiction, and Human Time: Historical Imagination and Historical Responsibility. In *The Ethics of History*, ed. David Carr, Thomas Flynn, and Rudolf Makkreel, 247–260. Evanston: Northwestern University Press.

Carrington, André M. 2016. *Speculative Blackness: The Future of Race in Science Fiction*. Minneapolis: University of Minnesota Press.

Chiang, Ted. 2002. *Stories of Your Life and Others*. New York: Vintage Books.
Clarke, Arthur C. 1973. *Rendezvous with Rama*. London: Gollancz.
El-Mohtar, Amal, and Max Gladstone. 2019. *This is How You Lose the Time War*. New York: Simon and Schuster.
Evans, Richard J. 2013. *Altered Pasts: Counterfactuals in History*. Waltham, MA: Brandeis University Press.
Fields, Karen E., and Barbara J. Fields. 2012. *Racecraft: The Soul of Inequality in American Life*. London: Verso.
Fur, Gunlög. 2017. Concurrences as a Methodology for Discerning Concurrent Histories. In *Concurrent Imaginaries, Postcolonial Worlds: Toward Revised Histories*, ed. Diana Brydon, Peter Forsgren, and Gunlög Fur, 3–32. Leiden: Brill.
Gallagher, Catherine. 2018. *Telling it like it wasn't: The Counterfactual Imagination in History and Fiction*. Chicago: University of Chicago Press.
Hennessey, John L. 2018. *Rule by Association: Japan in the Global Trans-imperial Culture, 1868–1912*. PhD diss., Linnaeus University.
Langer, Jessica. 2011. *Postcolonialism and Science Fiction*. New York: Palgrave Macmillan.
Lavender, Isiah. 2019. *Afrofuturism Rising: The Literary Prehistory of a Movement*. Columbus, OH: Ohio State University Press.
LeGuin, Ursula. 1971. *The Lathe of Heaven*. New York: Avon Books.
Lem, Stanisław. 1961. *Solaris*. Warsaw: MON.
Liedl, Janice. 2015. Tales of Futures Past: Science Fiction as a Historical Genre. *Rethinking History* 19 (2): 285–299.
Luckhurst, Roger. 2018. *Science Fiction: A Literary History*. London: British Library Publishing.
Macaulay, David. 1979. *Motel of the Mysteries*. New York: Houghton Mifflin.
Martine, Arkady. 2019. *A Memory Called Empire*. New York: Tor Books.
———. 2021. *A Desolation Called Peace*. New York: Tor Books.
Matsuda, Kyōko. 2003. *Teikoku no shisen: Hakurankai to ibunka hyōshō. [Imperial Gaze: Expositions and the Representation of Foreign Cultures]*. Tokyo: Yoshikawa Kōbunkan.
Mehan, Uppinder. 2004. Final Thoughts. In *So Long been Dreaming: Postcolonial Science Fiction & Fantasy*, ed. Nalo Hopkinson and Uppinder Mehan, 269–270. Vancouver: Arsenal Pulp Press.
Miller, Walter M. 1959. *A Canticle for Leibowitz*. Philadelphia: Lippincott.
Okorafor, Nnedi. 2015. *Binti: The Complete Trilogy*. New York: Daw Books.
Olsson, Anders. 2022. 15 miljoner kronor till forskning om urfolks överlevnad i Norden och Nordamerika [15 million SEK to Research on Indigenous Survival in the Nordic Countries and North America]. *Linnaeus University Website*, November 2. https://lnu.se/mot-linneuniversitetet/aktuellt/nyheter/2022/15-miljoner-kronor-till-forskning-om-urfolks-overlevnad

Painter, Nell Irvin. 2010. *The History of White People*. New York: Norton.
Pratt, Mary Louise. 1992. *Imperial Eyes: Travel Writing and Transculturation*. London: Routledge.
Rieder, John. 2008. *Colonialism and the Emergence of Science Fiction*. Middletown, CT: Wesleyan University Press.
Robinson, Kim Stanley. 2020. *The Ministry for the Future*. London: Orbit Books.
Rosenfeld, Gavriel D. 2005. *The World Hitler Never Made: Alternate History and the Memory of Nazism*. Cambridge: Cambridge University Press.
Seed, David. 2011. *Science Fiction: A Very Short Introduction*. Oxford: Oxford University Press.
Strugatsky, Arkady, and Boris Strugatsky. 2012 [1972]. *Roadside Picnic*. Translated by Olena Bormashenko. London: Gollancz.
Tchaikovsky, Adrian. 2015. *Children of Time*. New York: Tor Books.
———. 2019. *Children of Ruin*. New York: Tor Books.
Wachowskis, The, dir. 1999. *The Matrix*. Warner Bros.
Varnava, Andrekos, ed. 2015. *Imperial Expectations and Realities: El Dorados, Utopias and Dystopias*. Manchester: Manchester University Press.
Villeneuve, Denis, dir. 2016. *Arrival*. Paramount Pictures.
Wells, H.G. 1895. *The Time Machine*. London: William Heinemann.
———. 1898. *The War of the Worlds*. London: William Heinemann.
Willis, Connie. 1992. *Doomsday Book*. New York: Bantam Spectra.
———. 1997. *To Say Nothing of the Dog: Or, How We Found the Bishop's Bird Stump at Last*. New York: Bantam Spectra.

PART I

Colonialism, Oppression and Concurrences

CHAPTER 2

Concurrences and the Planetary Emergency: Ursula K. Le Guin in the Capitalocene

Johan Höglund

"Recognizing concurrences is about respecting pluralism and expecting entanglements."
Gunlög Fur

"Love doesn't just sit there, like a stone, it has to be made, like bread; remade all the time, made new."
Ursula LeGuin

The past, as Gunlög Fur (2017) reminds us, is full of concurrent and conflicting voices. Of these voices, the one that rises out of European modernity and the colonial and capitalist projects that made this modernity possible has been particularly loud and influential. As a long and vital tradition of radical, feminist, and postcolonial scholarship has shown, history has not simply been written by and for the people that expanded out of Europe to colonize most of the rest of the planet; the very discipline of history became a mechanism for Eurocentric power, as argued by Dipesh Chakrabarty (2000) and Gurminder K. Bhambra (2007) in somewhat

J. Höglund (✉)
Linnaeus University, Småland, Sweden
e-mail: johan.hoglund@lnu.se

J. L. Hennessey (ed.), *History and Speculative Fiction*,
https://doi.org/10.1007/978-3-031-42235-5_2

different ways. In conversation with this scholarship, Fur has proposed the concept of *concurrences* as a sociological/historical methodology that directs attention towards the voices that Eurocentric, universalizing, grand history has systematically drowned out. When the historian recognizes the presence of a plurality of voices (from indigenous people, from women, from workers, from the poor) and begins to listen to these voices, not just the past but the very discipline of history appears as entangled, fractured, and impossible, but also as a multi-layered repository and a practice full of possibilities.

This realization is arguably crucial for how the past is narrated within the discipline of history, but it is equally important for what futures we can imagine. Going back in time to listen to the voices that describe immensely rich life worlds not premised on normative European conceptions of civilization, gender, sexuality, and skin color yields a better and much more complex understanding of the past, while at the same time making it possible to imagine futures other than those extrapolated from the present moment of violent neocolonial/capitalist world dominance. At a time when the planetary emergency is unfolding against increasing international conflict and war, the ability to imagine new and different futures beyond those seemingly programmed into what Immanuel Wallerstein (2004) has termed the world-system, is more important than ever. As scholars such as Donna Haraway (2015) and Jason W. Moore (2015) have convincingly argued, colonialism and capitalism (two systems so entangled that they cannot be usefully separated) have reorganized human society and ecology since the early colonial period and set the planet towards the present moment of ecological, environmental collapse. Focusing on the role that the slave plantation, as a site of white, heteronormative exploitation and extraction, Haraway has suggested the concept *Plantationocene* as an alternative to the more popular moniker Anthropocene. Centering the same colonial and capitalist history, Moore has proposed the alternative term *Capitalocene*. Both concepts recognize how the present moment of climate upheaval and global ecological and economic injustice grows out of a specific, post-1500, capitalist and colonial history.

What can humans (and the various species with which humans are entangled) do to break out of a 500-year tradition of exploitation and extraction? What can humans do to halt this accelerating journey towards climate collapse, the sixth mass extinction of species (Kolbert 2014), and the oblivion that this is bringing about? What are the alternatives to (ever-expanding) capitalism, to violent geopolitics, to neoliberal globalization, and to the global social and ecological injustice that these systems underwrite? How do you put an end to the Plantationocene/Capitalocene? Arguably, a first step is to recognize and listen to the concurrent voices speaking (to us, with us, for us) from positions that may have been invaded by, but are not necessarily premised on, capitalism. The concurrent and complex political and social worlds that these voices rise out of can potentially help us imagine, and thus work towards, just and egalitarian futures not built on competition, exploitation, extraction, and accumulation.

Ursula Le Guin is an author who has always attempted to think outside the Eurocentric, Enlightenment paradigm and who has made the effort to imagine future worlds premised on social systems very different from those that fuel and organize present capitalist society. This chapter explores her novel *The Lathe of Heaven* (1971) as a narrative that identifies militarized capitalism and colonialism as the actual drivers of climate change and pandemic violence. The novel also, I argue, clearly traces capitalism and colonialism back to the origins of European Enlightenment. As I will argue, *The Lathe of Heaven* thus envisions a future where global warming caused by war and capitalism has brought the ecosystem to its knees. Oceanification, constant warfare, racism, a multitude of pandemics, and uneven and endemic precarity characterize the world in ways that are often disturbingly like the present moment and that square remarkably well with the Capitalocene or Plantationocene theses. While this alone makes *The Lathe of Heaven* into a singularly prescient novel, what makes it unique is its strangely optimistic vision of a multitude of concurrent futures—at once dystopian and utopian—through which the climate emergency, the precarity, the pandemics, and the violent conflict described in the novel can be undone or at least addressed.

Le Guin and the Capitalocene

To appreciate the original and complex account of the planet's environmental and political history provided by *The Lathe of Heaven*, it is necessary to, in more detail, consider recent scholarship that describes the forces that have produced the current emergency. This is in itself a fraught historical landscape of conflicting and competing voices and claims. In 2001, Paul J. Crutzen and Eugene F. Stoermer launched the concept of the Anthropocene to describe an epoch where human activity is the most substantial environmental and geological factor. While this puts the blame for the planetary emergency on the Anthropos, it arguably fails to separate the people who created the systems and technologies that actually produce climate change from indigenous and precarious communities that have done little to cause the emergency but that suffer the most from the deterioration of the Earth System. In other words, as influentially argued by sociologist John Bellamy Foster (Foster et al. 2011; Foster 2022) and by Moore (Moore 2015), it is capitalism, rather than the human as a species, that causes climate change. According to his scholarship, the beginning of the climate emergency is thus not to be found in human evolution, as the concept Anthropocene suggests, but in early capitalism and colonialism and in the Enlightenment project that legitimized and rationalized these systems.

These different versions of the origins and drivers of the planetary emergency have enormous significance for how this emergency can be resolved and narrated. In geology, environmental science, and environmental humanities, there is a clear tension between research and scholarship that sees the human species as the entity that has caused the climate emergency (Chakrabarty 2009; Steffen et al. 2016) and that identifies capitalism as the engine of the planetary emergency (Moore 2015; Malm 2016; Lewis and Maslin 2015; Davis and Todd 2017; Holleman 2018). If, as Chakrabarty has argued, the human species has become "a geological agent" (2009: 218), and if, as Steffen, Crutzen, and McNeill have suggested, the proper beginning of the climate emergency is the discovery of fire by Homo erectus "a couple of million years ago" (2016: 614), it becomes necessary to transform the species itself to halt the current development. If, however, as Moore, Malm, Lewis and Maslin, Davis and Todd, and Hollemann argue, it is extractive capitalism and colonialism that have caused the ongoing crisis, we do not have to stop being human in the sense of being *Homo sapiens*. Rather, we need to identify and embrace other ways of being human, other ways of being on this planet. This is not a simple task, but it is one that was once performed by millions of people for tens of thousands of years before the advent of European modernity.

Le Guin's substantial and radical science fiction oeuvre has been dedicated to interrogating this precise historical development, the gendered and racialized categories that energized this history, and the consequences for human society and for ecology that it produced. Entering the literary scene in the late 1950s, Le Guin began changing the way that American science fiction was written. Novels such as *The Left Hand of Darkness* (1969), *The Word for World is Forest* (1972), and *The Dispossessed* (1974) show a profound awareness of the United States' neocolonial project in

Southeast Asia, of the apartheid order practiced within the nation's borders, and of predatory capitalism in South America. In this way, Le Guin's writing has always explored many of the issues central to postcolonial studies. As Wendy Gay Pearson observes, novels such as *The Left Hand of Darkness* may not be "directly informed by postcolonial theory," but they are certainly "informed by the very conditions and historical circumstances that created both the postcolonial condition and the theory that attempts to explain and understand it" (189). Similarly, Le Guin was singularly aware of the havoc produced by patriarchal and heteronormative sexual orders and of the connection between these orders and the capitalist and colonial projects she so often critiqued in her writing. Thus, while *The Word for World is Forest* focuses on slavery and the ruthless sexual and ecological extractive violence practiced by capitalist imperialism, *Left Hand of Darkness* investigates and problematizes the connection between what Adrienne Rich (1980) has termed "compulsory heterosexuality" and (geo)political power. In the words of Pearson (2007), *The Left Hand of Darkness* may not have been the first science fiction novel to discuss colonialism, but it "can certainly be understood as central to any genealogy of works that link issues of gender and race to the history and legacy of colonialism" (184). In this way, as Sean McCann and Michael Szalay (2005) note, authors such as Don DeLillo and Toni Morrison can be said to "have followed the path charted by Le Guin" in the 1960s and 70s (447).

Le Guin has also been noted for using science fiction not simply to identify, allegorize, and decry unjust sexual, economic, and ecological relations but also to imagine alternative forms of social organization. Thus, Le Guin importantly refuses to see colonialism and capitalism as monolithic and inevitable. Indeed, as Carl D. Malmgren (1998) is one of many to note, many of Le Guin's novels "deal philosophically with the idea of utopia, the possibility of a perfect society" (313). Le Guin's writing rarely, if ever, takes the reader on a journey that ends in the triumphant establishment of such a utopia. Instead, her fiction inventories a range of possible relations between peoples, societies, ecologies, sexualities, and cultures that together make it possible to both perceive dominant capitalist and neocolonial society in a new light and to imagine various alternatives to capitalism's world order. While these futures are never untouched by capitalism and colonialism, they also transcend them by connecting to non-Western and indigenous cosmologies that refuse the extractive priorities that structure capitalist society. Drawing from Taoism (Galbreath 1980), pacifist anarchism (Call 2007), and indigenous systems of knowledge (Spicer 2021), Le Guin helps imagine future worlds beyond the one currently eroding conditions for life on the planet.

The Lathe of Heaven

The *Lathe of Heaven* is one of Le Guin's most intriguing and complex examples of such imagining. As I will argue, the novel attempts to explore the violent material dynamics of capitalist empire and the ways of thinking that have brought the world to its current state, but it also explores alternatives to this world order. The text is set in a dystopian Portland in 2002, some 30 years into the future counting from the date it was published. Social injustice is rampant in this dark future: "Undernourishment, overcrowding, and pervading foulness of the environment were the norm. There was more scurvy, typhus, and hepatitis in the Old Cities, more gang violence, crime, and murder in the New Cities" (28). In addition to this, the "Greenhouse Effect" has melted the snow from "all the world's mountains" (7) and turned the Portland sky a constant gray. Importantly, Le Guin attributes this not to humanity as a species but to the systems that keep capitalist society running:

> Very little light and air got down to street level; what there was was warm and full of fine rain. Rain was an old Portland tradition, but the warmth—70 °F on the second of March—was modern, a result of air pollution. Urban and industrial effluvia had not been controlled soon enough to reverse the cumulative trends already at work in the mid-twentieth century; it would take several centuries for the CO_2 to clear out of the air, if it ever did. (27)

The process that has led to the collapse of the climate is implicitly linked to other types of (military) violence. The first section of the novel takes place against a background of constant geopolitical conflict. In a striking passage, considering the fact that the novel is set in 2002, the reader is confronted with a newspaper headline that reads: "BIG A-l STRIKE NEAR AFGHAN BORDER [...] Threat of Afghan Intervention" (27). As in the present moment, much of the conflict occurs in or around the oil-rich Middle East.

In this uncomfortably prescient setting, a man named George Orr illegally collects chemical stimulants capable of keeping him awake for long periods of time. Discovered by police and health services, he is referred to the claustrophobic, windowless office of psychiatrist Dr. William Haber for "Voluntary Therapeutic Treatment" (8). When Haber begins to question Orr, the latter informs the psychiatrist that he has "effective dreams": "dreams that … that affected the … non-dream world. The real world" (11). Haber is predictably skeptical of this claim, but he discovers, after having sent Orr into a hypnotic sleep, that the assertion is true. Orr's dreams do come true. This is not noticed by other people in the world. To them, the new world is the only world there has ever been. Only Orr and, for some reason, Haber, while he makes Orr dream, remember what the world used to be like.

As critics have observed (Johnston 1999; Malmgren 1998), the name George Orr references George Orwell, another author of prescient, dystopian science fiction. Haber, meanwhile, connotes Homo Faber: a word describing, and simultaneously inventing, the human as a rational and productive species. Suspicious of any attempt to manipulate the future, Orr is extremely wary of his ability: "Who am I to meddle with the way things go? And it's my unconscious mind that changes things, without any intelligent control" (14). By contrast, Haber's rational persona is immensely stimulated by the possibilities that Orr's dreaming provides. With the help of hypnosis and a machine developed by "Russians" and "Israelis" that provides him with a modicum of control over Orr's dreams, Haber soon exits his cramped office by making Orr dream up the government-funded Oregon Oneirological Institute of which Haber is the director. Comfortably ensconced in this new and prestigious institution, he continues to manipulate the world he inhabits.

The utopian potential of such manipulation is compromised by the erratic behavior of dreams. When Haber makes Orr dream about the end of the overpopulation crisis, the result is a world where a plague has erased most of humanity. Suddenly, Orr's and Haber's minds are invaded with a new set of memories that vie with the old so that Haber now discovers new memories that tell him that there "are no floods now in the Ganges caused by the piling up of corpses of people dead of starvation. There's no protein deprivation and rickets among the working-class children of Portland, Oregon. As there was—before the Crash" (68). Thus, the end of overpopulation comes at a terrible price. Invaded by the memories this altered reality brings with it, Haber realizes that:

> I was already a grown man when the first epidemic struck. I was twenty-two when that first announcement was made in Russia, that chemical pollutants in the atmosphere were combining to form virulent carcinogens. The next

> night they released the hospital statistics from Mexico City. Then they figured out the incubation period, and everybody began counting. Waiting. And there were the riots, and the fuck-ins, and the Doomsday Band, and the Vigilantes. And my parents died that year. My wife the next year. My two sisters and their children after that. Everyone I knew. (68)

In this way, Orr and Haber find themselves having survived a catastrophic pandemic named "The Crash," and they now inhabit a phase known as "The Recovery," when the big cities on the U.S. East and West coasts are trying to amend the damage done not just by the plague but to ecology before the plague. Indeed, Orr notes how the "air was still profoundly and irremediably polluted: that pollution predated the Crash by decades, indeed was its direct cause" (80).

By connecting the pollution of the air by human systemic activity to the eruption of a devastating global pandemic, Le Guin's writing seems to speak directly to our own dystopian moment. A study published in *Nature Climate Change* in late 2022 reveals that climate change exacerbates 65 percent of all pathogenic diseases that affect humans (Mora et al. 2022), and a number of scholars in the environmental humanities and development studies (Malm 2020; Selby and Kagawa 2020; Duncan and Höglund 2021) have argued that the Covid-19 pandemic was provoked by, disseminated via, and experienced through the very same systems that are causing the climate emergency. It is the mapping of such connections that has led Jude Fernando (2020) to suggest that we now live in the "Virocene," an era where the detrimental and uneven ecological and economic conditions created by capitalism and colonialism will continue to produce pandemics.

Orr is deeply distressed by the transformations that Haber forces him to perform through the therapy sessions. "Please, stop using my dreams to improve things, Dr. Haber. It won't work. It's wrong. I want to be cured," (81) he pleads. But Haber remains unconcerned: "isn't that man's very purpose on earth—to do things, change things, run things, make a better world?" (82) he asks in reply to Orr's plea. Indeed, for Haber, some things have improved. Now that grain is not as precious as before the plague, there is suddenly bourbon, instead of rubbing alcohol, in his office desk and he confidently toasts the empty city outside his office: "To a better world!" (72). Never doubting that he is a "benevolent man" who wants to "make the world better for humanity" (83) he takes on the next challenge. The plague may have alleviated the population crisis, but it has done little to relieve international tension: "Jerusalem was rubble, and in

Saudi Arabia and Iraq the civilian population was living in burrows in the ground while tanks and planes sprayed fire in the air and cholera in the water, and babies crawled out of the burrows blinded by napalm" (81). Undaunted by the killing of billions caused by the depopulation dream, Haber sets out eradicate war and hypnotizes Orr to dream a dream about peace. The result is again unpredictable and violent. Orr dreams up a hostile alien species and while this ends war between humans on Earth, war is now moved into space, throwing humanity into yet another crisis.

Orr and Haber manage to dispel this crisis by redreaming the aliens into a generous and pacifist species that enters and interacts with the human population. While this suspends international and interplanetary war, in the process introducing a new species with their own set of creeds and cosmologies into human society, it does not put an end to Haber's futile attempts to design an enlightenment utopia. An effort to erase racism turns all people a shade of grey, social stability comes at the price of authoritarianism and blood sports, and the pandemic is addressed through a strict Eugenic regime where all people with "a serious communicable or hereditary disease" (135) are strictly monitored and euthanized on the spot when digressing from the harsh regime enforced upon them. Simultaneously with these developments, Haber continues to better his own situation. Towards the end of the narrative, Haber resides in an enormous building inspired by the Roman Pantheon and occupying an area larger than the British Museum. Over its entrance, the words "The Greatest Good for the Greatest Number" are written, and inside, the visitor is confronted with a plaque that reads: "The Proper Study of Mankind is Man • A. Pope • 1688 • 1744" (136). In this way, Le Guin makes perfectly clear how Haber's project is aligned with the Enlightenment attempt to reshape the world according to rational principles that simultaneously fueled and concealed, as Bhambra (2007) and Walter Mignolo (2011) have argued, the colonial and capitalist European project.

Haber's final intervention is to make Orr dream a dream where he confers his ability to dream effective dreams to Haber himself. Haber is certain that his own rational mind is a better agent for dramatic social and ecological intervention than Orr's resistant and anti-authoritarian subconscious. The results of this transfer of dream power are predictably chaotic and catastrophic. Haber's vision is ultimately empty and meaningless. As Malmgren (1998) suggests, William Haber in the end "becomes his name, an empty 'Will-I-Am,' a naked 'will to power' that feeds on itself" (315–6). Because Haber's dreaming lacks true content and is steered by his own desire for power, his dreaming causes the world to fold, melt, and become disconnected and meaningless:

> The funicular was crossing the river now, high above the water. But there was no water. The river had run dry. The bed of it lay cracked and oozing in the lights of the bridges, foul, full of grease and bones and lost tools and dying fish. The great ships lay careened and ruined by the towering, slimy docks.
>
> The buildings of downtown Portland, the Capital of the World, the high, new, handsome cubes of stone and glass interspersed with measured doses of green, the fortresses of Government—Research and Development, Communications, Industry, Economic Planning, Environmental Control—were melting. They were getting soggy and shaky, like Jell-O left out in the sun. The corners had already run down the sides, leaving great creamy smears. (171)

Orr, not present during Haber's dreaming, sees this transformation occurring and understands what is going on. He immediately rushes to Haber's grand office and manages to abort this collapsing dream at the last minute. The result, however, is a bricolage of the many previous realities dreamed up by Orr:

> The emptiness of Haber's being, the effective nightmare, radiating outward from the dreaming brain, had undone connections. The continuity that had always held between the worlds or timelines of Orr's dreaming had now been broken. Chaos had entered in. He had few and incoherent memories of this existence he was now in; almost all he knew came from the other memories, the other dreamtimes. (174)

Thus, the Earth ends up an incoherent yet organic jumble made up of a patchwork of all the previous realities: instead of a single utopian/dystopian future, the planet enters a state where a number of concurrent realities exists simultaneously. Orr moves through a city

> half wrecked and half transformed, a jumble and mess of grandiose plans and incomplete memories, swarmed like Bedlam; fires and insanities ran from house to house. And yet people went about their business as always: there were two men looting a jewelry shop, and past them came a woman who held her bawling, red-faced baby in her arms and walked purposefully home.
>
> Wherever home was. (175)

Haber's mind, his capacity to dream effectively, does not survive his failure and the chaotic unreality it has produced. In a permanent, catatonic state, he is confined to an asylum. Orr, by contrast, finds the new and divergent reality strangely livable. People are no longer uniformly grey, post-Enlightenment capitalism is still operational, but it coexists with a barter economy and with the enormous turtle-like aliens now both integrated into and apart from human society.

Conclusion: *Concurrences* and Le Guin's Dystopian Utopias

While *The Lathe of Heaven* has not received the same degree of attention as *The Left Hand of Darkness* or *The Dispossessed*, critics have long attempted to come to terms with its enigmatic ending. To Malmgren (1998), this ending suggests that "Utopia is just such a contradiction, a true science fantasy, something that neither reason nor imagination can bring into existence; Utopia is either the nightmare expression of a dangerous selfishness or merely an empty dream" (322). Reading the novel from a traditional, postcolonial, Bhabhian perspective and alongside Le Guin's *The Telling* (2000), Sahar Jamshidian and Farideh Pourgiv (2019) see in her oeuvre a general critique of "cultural imperialism" that "rejects both conservative and assimilative attitudes toward the other" and that "praises hybridity as the culture of our globalizing world" (96). These two readings recognize a prevalent tension in *The Lathe of Heaven* as well as in Le Guin's writing generally, but neither acknowledge the prescient critique of Capitalocene violence that saturates the novel nor the strange hope that resides in the similarly strange ending of the novel, with multiple pasts and realities existing concurrently.

Homi K. Bhabha's (1994) concepts of third space and hybridity, which Jamshidian and Pourgiv employ to dissect Le Guin's writing, assume a coming together of differing perspectives and modes of being. Out of the wildly divergent modes of thinking and discourses that characterize the meeting between colonizer and colonized—what Bhabha refers to as colonial ambivalence—a third space is formed, and within this space, different discourses and views merge and hybridize. To Jamshidian and Pourgiv, globalization, if done right, is a force that encourages the forming of such spaces and such mergers. However, Le Guin's fiction is clearly suspicious of globalization as a process. As suggested by a host of scholars, including Michael Hardt and Antonio Negri (2000), David M. Kotz (2002), David

Harvey (2002), and Richard Peet (2009), globalization is best understood as the process by which neoliberal capitalism expands across the planet. Its roots, as Wallerstein (2000) argues, go back to the 1450s, when the current world-system came into being. This beginning is also, of course, the beginning of what Moore has termed the Capitalocene. The origins of extractive capitalism and of European colonialism are thus also the origin of both globalization and of the planetary emergency (Moore 2015; Moore 2016). Le Guin is clearly aware of the existence of an extractive and violent capitalist and colonial world-system and of this system's capacity to generate ecocide, war, and plagues rather than third spaces where hybridity can erupt. It is thus not surprising that Haber's attempt to set the world right by relying on the same Enlightenment principles that underpin this system forces the world further into chaos. Orr's disruption of Haber's misguided and futile enlightenment dreaming does not open the door to a third space where hybridity can take place, but rather to a future where several dystopian, colonial, and capitalist worlds co-exist and vie for space with utopian hopes and projects.

In other words, the divergent ending of the novel does not evidence the type of hybridization that Bhabha suggests occurs in colonial and transcultural spaces. Rather, what emerges out of the chaos is an entangled state full of concurrent voices, of plural world views and practices. Unlike the voices that Fur's historical research has helped to bring to our attention, Le Guin's voices speak to us from an imagined future, yet these voices speak about similar issues: the violence performed on planets and people, the possibility of escaping or unthinking the racialization of people, and the existence of worlds and of modes of being detached from capital and patriarchy. Thus, considered through the methodological and theoretical lens that *concurrences* provides, the ending of *The Lathe of Heaven* cannot usefully be read as a coming together of a disparate chorus. Rather, as I have argued elsewhere, it exemplifies how literature often supplies the discordant hum of "a number of concurrent voices that speak simultaneously and outline multiple, sometimes wildly divergent positions" (Höglund 2013: 288). The sound of such a plethora of voices, many of them repeating the central tenets of capitalism, does not erase the history and ongoing violence of capitalist and colonial violence, nor does it elide the global warming, the social injustice and the pandemics that produce havoc in Le Guin's dystopian 2002, and that are doing so also in the present moment. But these processes now exist concurrently with other worlds and other modes of being. The has not been eliminated, but it exists alongside ways of interacting with the world that are not necessarily locked into its binaries and logic. People are grey, brown, and white, love is possible across these artificial divides and aliens the shape of giant turtles have opened second-hand shops in the street.

In contrast to Malmgren's analysis of *The Lathe of Heaven*, the realization that the ending of the novel contains a plethora of concurrent worlds, possibilities, and futures suggests that the novel does not dismiss the notion of utopia altogether. Rather, it envisions a world where alternatives to capitalism and colonialism—the—exist but where the history of this system is still in place. In an important article on indigenous science fiction, Kyle P. Whyte (2018) warns readers, critics, and authors of science fiction to avoid imagining indigenous people as "Holocene survivors" somehow unburdened by 500 years of genocide and colonialism. Like the ancestors of European settlers, indigenous people have been thoroughly transformed by 500 years of colonialism and extraction. Thus, indigenous science fiction does not (should not) pretend, like much other science fiction, that the ecological and social collapse the planetary emergency is producing is something new. From the perspective of a very long history of settler capitalist violence, the ongoing planetary emergency is part of a process that began with the arrival of Christopher Columbus, Ferdinand Magellan, and James Cook.

Although not an indigenous author, Le Guin is clearly aware, and deeply critical, of the material and epistemological violence that has long been practiced on indigenous people, on women, and on the poor. The wars, the illnesses, and the global warming that the novel depicts are intimately connected to the same Enlightenment history that includes both Columbus and Alexander Pope. The concurrent worlds that mark the end of the novel erase none of that violence. Capitalism and colonialism are less dominant in this fragmented world, but they remain in place, and they are growing. But the divergent worlds that exist at the end of the novel also make other futures possible. Trajectories can be discerned that open up futures not premised on capitalism or colonialism. Orr, inspired by conversations with the aliens he has dreamed up, says: "when the mind becomes conscious, when the rate of evolution speeds up, then you have to be careful. Careful of the world. You must learn the way. You must learn the skills, the art, the limits. A conscious mind must be part of the whole, intentionally and carefully" (167). This is a worlding and a form of being radically different from the mechanics of occupation, extraction, and consumption that through which the world-system was build, and it makes it possible to think of futures very different from those generated by this system. Indeed, (like love or like bread) the future is not a thing set in stone; it can be "remade all the time, made new" (159).

References

Bhabha, Homi K. 1994. *The Location of Culture*. London: Routledge.
Bhambra, Gurminder. 2007. *Rethinking Modernity: Postcolonialism and the Sociological Imagination*. Basingstoke: Palgrave Macmillan.
Call, Lewis. 2007. Postmodern Anarchism in the Novels of Ursula K. Le Guin. *SubStance* 36: 87–105.
Chakrabarty, Dipesh. 2000. *Provincializing Europe*. Princeton: Princeton University Press.
———. 2009. The Climate of History: Four Theses. *Critical Inquiry* 35: 197–222.
Davis, Heather, and Zoe Todd. 2017. On the Importance of a Date, or, Decolonizing the Anthropocene. *ACME: An International Journal for Critical Geographies* 16: 761–780.
Duncan, Rebecca, and Johan Höglund. 2021. Decolonising the COVID-19 Pandemic: On Being in this Together. *Approaching Religion* 11: 115–131. https://doi.org/10.30664/ar.107743.
Fernando, Jude L. 2020. The Virocene Epoch: The Vulnerability Nexus of Viruses, Capitalism and Racism. *Journal of Political Ecology* 27: 635–684. https://doi.org/10.2458/v27i1.23748.
Foster, John Bellamy. 2022. *Capitalism in the Anthropocene: Ecological Ruin or Ecological Revolution*. New York: NYU Press.
Foster, John Bellamy, Brett Clark, and Richard York. 2011. *The Ecological Rift: Capitalism's War on the Earth*. New York: NYU Press.
Fur, Gunlög. 2017. Concurrences as a Methodology for Discerning Concurrent Histories. In *Concurrent Imaginaries, Postcolonial Worlds: Towards Revised Histories*, ed. Diana Brydon, Peter Forsgren, and Gunlög Fur, 33–58. Leiden: Brill.
Galbreath, Robert. 1980. Taoist Magic in the Earthsea Trilogy. *Extrapolation* 21: 262.
Haraway, Donna. 2015. Anthropocene, Capitalocene, Plantationocene, Chthulucene: Making Kin. *Environmental Humanities* 6: 159–165.
Hardt, Michael, and Antonio Negri. 2000. *Empire*. Cambridge, MA: Harvard University Press.
Harvey, David. 2002. The Art of Rent: Globalisation, Monopoly and the Commodification of Culture. *Socialist Register* 38.
Höglund, Johan. 2013. Black Englishness and the Concurrent Voices of Richard Marsh in *The Surprising Husband*. *English Literature in Transition, 1880-1920* 56: 275–291.

Holleman, Hannah. 2018. *Dust Bowls of Empire: Imperialism, Environmental Politics, and the Injustice of "Green" Capitalism*. New Haven: Yale University Press.

Jamshidian, Sahar, and Farideh Pourgiv. 2019. Local Heritage/Global Forces: Hybrid Identities in Le Guin's *The Telling*. *Gema Online Journal of Language Studies* 19: 96.

Johnston, Laura. 1999. "Orr" and "Orwell": Le Guin's *The Lathe of Heaven* and Orwell's *Nineteen Eighty-Four*. *Extrapolation* 40: 351–351. https://doi.org/10.17576/gema-2019-1904-05.

Kolbert, Elizabeth. 2014. *The Sixth Extinction: An Unnatural History*. New York: Henry Holt.

Kotz, David M. 2002. Globalization and Neoliberalism. *Rethinking Marxism* 14: 64–79.

Le Guin, Ursula K. 1971. *The Lathe of Heaven*. New York: Scribner.

Lewis, Simon L., and Mark A. Maslin. 2015. Defining the Anthropocene. *Nature* 519: 171–180. https://doi.org/10.1038/nature14258.

Malm, Andreas. 2016. *Fossil Capital: The Rise of Steam Power and the Roots of Global Warming*. London: Verso Books.

———. 2020. *Corona, Climate, Chronic Emergency: War Communism in the Twenty-First Century*. London: Verso.

Malmgren, Carl D. 1998. Orr Else? The Protagonists of LeGuin's *The Lathe of Heaven*. *Journal of the Fantastic in the Arts* 9: 313–323.

McCann, Sean, and Michael Szalay. 2005. Do You Believe in Magic? Literary Thinking after the New Left. *The Yale Journal of Criticism* 18: 435–468.

Mignolo, Walter. 2011. *The Darker Side of Western Modernity: Global Futures, Decolonial Options*. Durham: Duke University Press.

Moore, Jason W. 2015. *Capitalism in the Web of Life: Ecology and the Accumulation of Capital*. London: Verso.

———. 2016. The Rise of Cheap Nature. In *Anthropocene or Capitalocene? Nature, History, and the Crisis of Capitalism*, ed. Jason W. Moore, 78–115. Oakland, CA: PM Press.

Mora, Camilo, Tristan McKenzie, Isabella M. Gaw, Jacqueline M. Dean, Hannah von Hammerstein, Tabatha A. Knudson, Renee O. Setter, Charlotte Z. Smith, Kira M. Webster, and Jonathan A. Patz. 2022. Over Half of Known Human Pathogenic Diseases can be Aggravated by Climate Change. *Nature Climate Change* 12: 869–887. https://doi.org/10.1038/s41558-022-01426-1.

Pearson, Wendy Gay. 2007. Postcolonialism/s, Gender/s, Sexuality/ies and the Legacy of *The Left Hand of Darkness*: Gwyneth Jones's Aleutians Talk Back. *The Yearbook of English Studies* 37: 182–196.

Peet, Richard. 2009. *Unholy Trinity: The IMF, World Bank and WTO*. London: Bloomsbury Publishing.

Rich, Adrienne. 1980. Compulsory Heterosexuality and Lesbian Existence. *Signs: Journal of Women in Culture and Society* 5: 631–660.

Selby, David, and Fumiyo Kagawa. 2020. Climate Change and Coronavirus: A Confluence of Two Emergencies as Learning and Teaching Challenge. *Policy & Practice: A Development Education Review* 30: 104–114.

Spicer, Arwen. 2021. Many Voices in the Household: Indigeneity and Utopia in Le Guin's Ekumen. In *The Legacies of Ursula K. Le Guin*, ed. Christopher L. Robinson, Sarah Bouttier, and Pierre-Louis Patoine, 65–81. Cham: Palgrave Macmillan.

Steffen, Will, Paul J. Crutzen, and John R. McNeill. 2016. The Anthropocene: Are Humans Now Overwhelming the Great Forces of Nature? In *The New World History*, ed. Ross E. Dunn, Laura J. Mitchell, and Kerry Ward, 440–458. Oakland: University of California Press.

Wallerstein, Immanuel. 2000. Globalization or the Age of Transition? A Long-Term View of the Trajectory of the World-System. *International Sociology* 15: 249–265.

———. 2004. *World-Systems Analysis: An Introduction.* Durham: Duke University Press.

Whyte, Kyle P. 2018. Indigenous Science (Fiction) for the Anthropocene: Ancestral Dystopias and Fantasies of Climate Change Crises. *Environment and Planning E: Nature and Space* 1: 224–242. https://doi.org/10.1177/2514848618777.

CHAPTER 3

Concurrent Whiteness: Science Fiction Film's Close Encounters in Apartheid South Africa

Ashleigh Harris

23 March 1979, South Africa. Less than three years earlier, the murder of schoolchildren during the Soweto uprisings of June 1976 had drawn the apartheid state under international scrutiny. Only 18 months before, in September 1977, Steve Biko's death whilst in police detention consolidated global solidarity against the country. In 1979, South Africa was in political turmoil: *Umkhonto we Sizwe*, the military wing of the African National Congress, was active with guerrilla attacks on South African strategic targets, particularly railway lines and police stations; South African troops were actively involved in conflict in Angola against the South-West Africa People's Organisation, a socialist organization with ties to the Soviet Union; and neighboring Zimbabwe created a transitional government ahead of its first democratic elections in 1980.

Amidst these swathes of grand history was a banal event of 23 March 1979: the premiere of Steven Spielberg's blockbuster science fiction film, *Close Encounters of the Third Kind* (1977; hereafter *CE3K*) in South African cinemas. The mundaneness of this event gives us pause when we stop to consider the concurrent realities occurring in the country at the

A. Harris (✉)
Uppsala University, Uppsala, Sweden
e-mail: ashleigh.harris@engelska.uu.se

J. L. Hennessey (ed.), *History and Speculative Fiction*,
https://doi.org/10.1007/978-3-031-42235-5_3

time. How did cultural behavior as superficial and banal as going to watch science fiction films occur in a country where, only a stone's throw away from whites-only cinemas, *Umkhonto we Sizwe* guerrillas were putting their lives on the line daily to bring down the apartheid state? In the chapter that follows, the incongruity of this concurrent experience of Black and white life in South Africa during apartheid is extended to a broader scope in order to read the complicities of global cultural networks in their concurrence with South African apartheid.

In "Concurrences as a Methodology for Discerning Concurrent Histories," Gunlög Fur elaborates the term "concurrences" as a concept that provides the "methodological and theoretical foundations for studies that allow multiple voices to be heard; stories that voice concurrent claims on geographical, temporal, political, and moral spaces" (Fur 2017, 40). What concurrences enables is simultaneous, if at times contradictory and conflicting, accounts of history to be placed in productive conversation with one another. While the concept lends itself specifically to subaltern voicings of suppressed histories, in this chapter, I wish to explore the relevance of the term for understanding the global reach of a violently dominant discourse: that of whiteness under South African apartheid of the 1980s. This investigation is prompted by Fur's further observation that "[e]ngaging with 'the global' in whatever form it takes poses challenges [...] to seek ways to interpret worldwide processes and interactions and to ask to what degree global perspectives can and will grapple with a legacy of universalizing expressions and underlying claims centred in the West" (Fur 2017, 37). I posit that South African whiteness during apartheid, which was overtly militarized, politicized, and violent, was—in its most banal and everyday forms—mirrored, sustained, and even confirmed by contemporaneous Western popular culture.

Following Hannah Arendt's observation in *Eichmann in Jerusalem* that the "trouble with Eichmann was precisely that so many were like him, and that the many were neither perverted nor sadistic, that they were, and still are, terribly and terrifyingly normal" (2006, 274), this analysis looks at how the global complicities of whiteness—here via popular culture—enabled a banal maintenance of apartheid ideology in South Africa itself. This observation runs counter to a significant body of media scholarship that has pointed out the importance of global media and culture in shifting white South Africa's commitment to the racist state (see Hyslop 2000; Nixon 1994). My argument here is that there were concurrent reiterations of racist and racialized discourse flowing through global channels that affirmed, rather than critiqued, South African white self-perception. In the spirit of the method of concurrent readings, this does not undermine or make a claim for greater authenticity over the analyses that see popular culture as one of the mediators of change in South Africa. Rather, this chapter aims to thicken that account by considering South African whiteness not in a state of exception or pure isolation but rather as in complex dialogue with global discourses of race and whiteness. To put it simply: South African white everyday life was not far removed from white life elsewhere during the 70s and 80s but rather intimately connected and entangled with it.

This question matters because global confirmation of the claims of white life through popular culture worked as an enabler of white South African blindness to the violence that was sustaining their everyday, middle-class, privileged lives. Indeed, one of the most alarming concurrences of white life in Southern Africa in the 1970s and 1980s was that while civil unrest and state violence had reached unprecedented proportions during these years, white urban South African life continued largely unperturbed by what was happening in the townships. The notorious Group Areas Act of 1950, which demarcated residential zones according to ethnicity, was the crucial piece of apartheid legislation that enabled this experience of parallel life in South Africa. The reality of Black life under apartheid was literally lived at a remove from white life. This is not to say that Black South Africans did not exist for whites at the time: on the contrary, Black South Africans constituted the country's work force, from industry and mining to the intimacy of the domestic sphere of the home. But these workers were not transgressing the walls that apartheid established. Black life in the sphere of the white everyday under apartheid was what Giorgio Agamben has called "bare life," and a host of apartheid legislation ensured that, even when forced to be in the same locality, Black and white South Africans would never share—or equally participate—in a fully human life together.

Thus, while many commentators have noted the important role that increasing access to international media, television, film, and music in the country had on changing white perceptions of the Nationalist Party and its system of apartheid, I argue that such popular culture also operated as a kind of justification to maintain the status quo that secured middle-class white life at the time. The white, middle-class suburban lives lived out in American television programs, for example, confirmed white South Africans' sense of belonging to a global community. Indeed, at this time, white South Africa obsessively orientated its cultural identity towards a projected and implied version of white life in the global north. Jonathan Hyslop has argued that this, along with the emergence of a "massive Afrikaner middle class and a strong and confident business class" (Hyslop 2000, 38) in the beginning of the 1970s, created a new consumerist Afrikaner identity with a greater investment in personal autonomy. Ultimately, this new consumer class "wanted to separate their Afrikanerness from the history of race conflict and to attain international acceptance as 'modern' people" (Hyslop 2000, 41). That international acceptance was largely formulated around consumerist *participation* in global culture, such as television and film, sports, popular music, and global media events.

In *Broadcasting the End of Apartheid*, Martha Evans argues that exclusion from global events was as determining for South African white identity as inclusion in them, since "the pleasure associated with media events converts into displeasure at exclusion from such events, and [...] this was one of the contributing factors leading to white acceptance of reform in South Africa" (Evans 2014, 4). Evans discusses local responses to South Africa's exclusion from the televised moon-landing to elaborate the significance of cultural exclusion for white self-perception. In 1969 in South Africa, however, television was still being held off by a government anxious about what exposure to international media would mean for (particularly English-speaking) white awareness of global critique of the apartheid system, and so South Africans could only listen to the radio transmission of the event. Evans notes that this "exclusion from this pinnacle of human achievement coincided with a more cosmopolitan and progressive sense of identity," (32) which amplified white South African "fear of being perceived as 'backward'" (33). Rob Nixon makes the same point when he writes that

> For many whites—already rendered paranoid by the growing force of their exile from world affairs—South Africa's inability to partake of such a singular moment of "global" community came to seem like an exasperating self-inflicted disinvitation. A *Rand Daily Mail* editorial captured this sense of let-down perfectly with the snappy headline "Out of this World". (Nixon 1994, 74, cited in Evans 2014, 33)

The *Rand Daily Mail's* headline, "Out of this World" captures, too, the global stakes of the space race in cold war politics, which is a factor I will return to in discussing the later relevance of science fiction film in South Africa in the 1970s and 80s.

What all of this amounts to is that white South Africans' investment in media and popular culture from the global north was fueled by an anxiety about exclusion from a global community.[1] It is therefore not surprising that the international anti-apartheid movement pressed so hard for a cultural boycott on South Africa during the 1980s. But this aim to isolate the oppressive state through cultural boycott was far from successful: even when European countries or individual US music or film companies joined the boycott, popular culture still enjoyed relatively free (albeit sometimes pirated) circulation with the help of the contemporaneous technologies of the portable cassette and VHS recording devices. In the music world, when pirating was not an option, popular international songs that had not come to South Africa because of boycotts were recorded by local artists as cover versions and sold as compilation LPs (known as the Springbok Hit Parade albums). In some cases, this created a literal white-washing of songs by Black musicians, since the music of performers such as Michael Jackson and Stevie Wonder were rerecorded by white South African performers. This culture of emulation was significant in relation to television too, as seen in another major media event discussed by Evans: the Royal Wedding of 1981. The British Equity ban "banned exports to South Africa of all recorded material involving British Equity members" (Evans 2014, 45). The South African Broadcasting Corporation (SABC), however, "managed to find their way around the ban by importing programmes from other countries and even by adapting British programmes, for example the animated children's programme, *Rupert the Bear*. To get around the ban, the SABC dubbed the programme *from* English *into* English, as it featured performances by Equity voice artists" (Bevan 2008, 167; discussed by Evans 2014, 45, emphasis in the original). The Royal Wedding involved performances by numerous British Equity members, and as such, while "the footage of the ceremony was shown, the music played by Equity members was withheld—a problem that the SABC, by now adept at the practice of dubbing, easily circumvented by playing prerecorded music performed by non-Equity members" (Evans 2014, 48).

The practice of dubbing was indeed not only a way of maintaining the government's demand for an Afrikaans quotient on television, thereby aiding the apartheid culture industry, but also a way to remediate the flows of global culture, sometimes reworking them entirely. A fascinating example is discussed by Cobus van Staden in his discussion of the dubbing into

Afrikaans of a Japanese animated version of Finnish children's writer Tove Jansson's Moomin novels. Van Staden writes:

> Quantum Productions, the dubbing house, did not receive the customary full script in German, French or English but only the Japanese original, with a simple skeleton outline for each episode. While several of the scriptwriters were fluent in German and French, none of them knew any Japanese. They also did not have the time to consult the original *Moomintroll* novels. The production team therefore used the skeleton outlines they received to write new dialogue and even subplots to fit the animation. In addition, they renamed the characters and invented their motivations, personalities and modes of expression from scratch. (Van Staden 2014, 6)

This amounted to a major rewriting of the animated series to the extent that, as Van Staden notes, "the character of Too-Tickey (Tjoek-Tjoek in Afrikaans) [who] is female in the original novels as well as in the Japanese version, [became] male in the Afrikaans version, due to her short hair" (Van Staden 2014, 7).

The tenacity of white South Africa to keep up-to-date with Western popular culture ensured that these alternative routes of cultural circulation remained open. These examples complicate the idea that "Television programmes reflecting the relatively liberal ethos of the 1970s in America were dramatically out of kilter with prevailing white South African representations of race, gender and sexuality" (Hyslop 2000, 39). While this was certainly part of the story, I will focus on the concurrence or simultaneity of global culture and its South African mediations to show how particularly U.S. popular culture both challenged and upheld racial representations in South Africa. To this extent, I agree with Hyslop's argument that "[c]onfronted with a new set of racial representations, white viewers somehow had to reconcile them with their pre-existing conceptions. And at the same time as white South Africans were absorbing and internalizing the consumerist values of the American soap operas, during the 1980s, television was also bringing them news of their increasing rejection by the US, as support for sanctions mounted" (Hyslop 2000, 39). But the extent to which these cultural products echoed local racist conceptions of blackness and whiteness and the extent to which the apartheid cultural industry mediated viewers' experiences of these products[2] suggests that the concurrence of whiteness in popular culture in South Africa was a far more complex matter than Hyslop's argument accounts for.

Thus, despite international calls for the cultural isolation of the country, international popular culture flowed relatively freely, if mediated in various ways, in South Africa, and in turn played a crucial role in mediating white projections of what constituted European and American life. This global concurrence of white life—facilitated by popular culture—fractures the idea that apartheid was a limited, geographically and politically contained and isolated event. To articulate this idea through a theoretical register, we can turn to Jacques Derrida, who in his preface to *Spectres of Marx* (which was dedicated to South African communist party leader, Chris Hani)[3] touches on the global dimensions of apartheid (Derrida 1994). As Monica Popescu has pointed out, Derrida—articulating a philosophy of global responsibility—viewed events taking place in apartheid South Africa as standing "in metonymic relation to those in the world as a whole" (Popescu 2007, 2). Derrida writes: "At once part, cause, effect, example, what is happening there translates what takes place here, always here, wherever one is and wherever one looks, closest to home" (1994, xv). Derrida's collapsing of geo-political space insists on a sharpening of global responsibility. While the anti-apartheid solidarity movement certainly felt a sense of responsibility to respond to the immorality of what was happening in South Africa, Derrida's version of responsibility was never fully potentiated in that movement that tended, on the whole, to see the South African state as an exceptional one, rather than one that was sustained by international relations, as well as cultural and ideological resonances. In the analysis that follows, I attempt to potentiate Derrida's metonymic reading of South African apartheid by considering the ways in which apartheid, as a political and cultural formation, was folded into its concurrent global context. I will, thus, be reading whiteness in apartheid South Africa as concurrent, rather than isolated, as resonant with, rather than radically different to, global forms of whiteness that dominated popular culture at that time.

An Incomplete Isolation: Censorship and the Cultural Boycott

There were two processes driving South African isolation under apartheid. The first was the apartheid state's paranoia about the transformative—even revolutionary—potential of the media, as reflected in the state's censorship board; and the second was the increasing intensity of the call for a cultural boycott and economic sanctions by the international anti-apartheid solidarity movement. The state's paranoia about mass media is nowhere more evident than in the following statement by the minister of defense, Magnus Malan, in 1981:

> The primary aim of the enemy is to unnerve through maximum publicity. In this regard we will have to obtain the co-operation of the South African media in not giving excessive and unjustified publicity to the terrorists and thus playing into their hands. (cited in Tomaselli 1988, 20)

This anxiety about the media's complicity with the "enemy" was part of a longstanding wariness of the apartheid state towards mass media. A pertinent example is, as mentioned above, the state's refusal to implement television in the country until as late as 1976, making South Africa one of the last countries in the world to get TV. As television scholar Ron Krabill notes, the denial of television as an "…explicit attempt on the part of the apartheid regime to resist transnational media flows—particularly representations of the civil rights movement in the United States—… exacerbated many White South Africans' (particularly English speakers') sense of exclusion from the international community" (Krabill 2010, 11–12). Thus, by the time that television was introduced, it became invested by white South Africans with a sense of global participation. Given the severe economic imbalance that marked South African social life in the 1970s and 80s, the projected audience of television was for the most part white: a point reinforced by the fact that "as late as 1986 approximately three-quarters of television viewers were White South Africans" (Krabill 2010, 6), and the fact that when television started, it was broadcast only in Afrikaans and English and in white urban areas (Evans 2014, 37).

Despite the various mediations discussed earlier, television in South Africa in the 70s and 80s was the standard fare of, largely, US sitcoms and series enjoyed all over the English-speaking world. During the years of South Africa's increasing global disrepute, white South Africans were coming home from school and work to shows like *Dallas*, *Dynasty*, *Silver Spoons*, *Growing Pains*, *Family Ties*, and *Who's the Boss*; to action dramas like *The A-Team*, *Airwolf*, *Macgyver*, *Knight Rider*, *Magnum P.I.*, and *Miami Vice*; to sci-fi series like *Star Trek*, *Mork and Mindy*, *V*, *Tales from the Dark Side*; and to a host of children's animated series based on European classics.[4] Because the apartheid state was deeply anxious about the extent to which alternative racial forms of social organization might appeal to white South African audiences, it saw the representational sphere of popular culture as a significant threat in this regard. This anxiety had a longer legal history in South Africa, where, as early as 1931, film scenes representing the "intermingling between Europeans and non-Europeans" were banned (Tomaselli 1988, 14). This was amended in 1974, where the racial basis of censorship was cancelled, but the right to restrict films "to persons in a specific category…or at a specific place" (Tomaselli 1988, 24) was retained. Thus, Black South Africans would not see films representing the civil rights movement, for example. Also, content representing sexual relationships across the color bar was still banned, as were shows/films that were overtly critical of the apartheid state.

Interestingly enough, this flood of popular television that graced South African screens in the 1970s and 80s, all of which was carefully vetted by the state before being cleared for airing, seldom transgressed even the draconian laws of 1931. The state's paranoia towards popular culture is clear in the 1974 amendment to the censorship act, where we read: "The more popular the material, the more likely it is to be undesirable" (from "Guidelines with regard to section 47 (2) Act 42 of 1974. Addendum to Appeal Board Case 43/82," cited in Tomaselli 1988, 24).[5] Yet, despite this anxiety about the popular, it is extraordinary to note that most of the content on white television and film theaters in South Africa under apartheid actually met with the censorship board's approval. The first dimension of this observation is that there was a significant amount of television content coming into the country that confirmed a structured absence of Black life. The second dimension, as I try to illustrate in a close-reading of *CE3K* below, takes this structured absence as a given, but also suggests that we must read South African reception of this popular culture in ways that were specific to that particular time and location.

One example of how popular culture became refracted by the prism of apartheid can be elaborated through considering some of the seeming contradictions to the structured absence of Blackness on South African television—think, for example, of *Webster*, *Miami Vice*, *The A-Team*, and *The Cosby Show*, all of which had Black protagonists that made overt in-character statements resisting apartheid ideology. Ron Krabill observes that one of the most notable contradictions was that "at the height of apartheid's States of Emergency in the mid-1980s, the most popular television show among White South Africans was *The Cosby Show*" (Krabill 2010, 1). Krabill's discussion covers the contradictions inherent in this fact, most significantly the absurdity of the fact that because images of

Nelson Mandela were banned at this time, "most South Africans could recognize Bill Cosby's face, but almost no one knew what the figurehead of the anti-apartheid struggle actually looked like" (Krabill 2010, 1).

For Krabill, the popularity of *The Cosby Show* speaks to the transformative potential of the medium of television during this time. He writes that

> First, television made possible a transformation in the subjectivities and identifications of White South Africans, and this transformation in turn altered the nature of politics in late-apartheid South Africa and continues to reverberate through current efforts towards democratic consolidation. Second, television served as an initial site of negotiation in which the absence of Black South Africans in public life was first dismantled, thus allowing for White South Africans to imagine themselves as part of the same polity with Black South Africans long before the formal inclusion of Black South Africans in institutional practice. (Krabill 2010, 5)

This point is significant but should be tempered by a second factor: that is, white South Africa's almost obsessive orientation of its cultural identity *away from Africa* and towards the global north. It was this cultural orientation towards the global north that created the largest fracture in what apartheid Prime Minister P.W. Botha called South Africa's "total strategy" to meet the "total onslaught" by an international community increasingly critical of apartheid (cited in Tomaselli 1988, 20). What Botha and others did not fully account for was the extent to which white South Africa's desire to be included in an international community might undermine—albeit inadvertently—its commitment to the project of apartheid (as argued by Evans 2014; Hyslop 2000; Krabill 2010; Nixon 1994).

While Krabill argues, then, that American television allowed white South Africans to "appropriate the language and attitude of 'racial tolerance'" (2010, 13), I would argue that a more obvious point of identification was these shows' popularity in the U.S. Key to my point here is the fact that this need to identify with "the people of the world"—as a project of performing a distinctly non-African modernity—was not always at odds with the racist tenets of apartheid that construed Black life as "pre-modern" and saw whites as the bearers of modernity to Africa. This is where we start seeing the consequences of the *mise en abyme* of global whiteness: South African whiteness might have imagined transformative relations across the color bar because of American television, but it did so in part to articulate its distance from its own historical situation and to establish a troubling distinction between African Blackness (posited as pre-modern) and American/European Blackness (posited as "civilized"). As one (anonymous) commentator puts it in Krabill's book: "I don't think that *The Cosby Show* was in any way subversive of late-apartheid ideology. I don't think it made the government of P.W. Botha or the SABC in the least bit uncomfortable. The Huxtables—so charming and approachable, so respectably middle class, so segregated (very few Whites in the show to my recollection, the Huxtable kids going to black colleges)—sat very well with how White South Africa wanted to see Black people.... Theo's ['Free Nelson Mandela'] poster[6] was a small price to pay for these messages" (Krabill 2010, 104).

Complicity and Concurrence

Before embarking on my reading of how *CE3K* accrued meaning in its reception in apartheid South Africa, I wish to distinguish between complicity, which I see as an active participation in the continuation of the apartheid regime through economic or political support, and concurrence of whiteness, as I am using it here.

To begin with complicity: the United Nations committed itself early to a cultural boycott on South Africa, requesting in 1968 that "All states and organizations suspend cultural, educational, sporting, and other exchanges with the racist regime and with other organizations or institutions in South Africa which practice apartheid" (U.N. Resolution 2396, adopted by the General Assembly on December 2, 1968; cited in Beaubien 1982, 7). Yet, as Camille Bratton (1977) and Michael Beaubien (1982) establish, in the entire voting history of the United States on the question of sanctions and boycott, the country was singular in its refusal to back calls for sanctions against South Africa during the 1970s and 80s. Gail Ann Reed further notes: "It was the only country that did not vote 'yes' on a single southern Africa resolution put to a roll-call vote" (Reed 1977 in Beaubien 1982, 8). This voting record is less surprising when one considers American economic interest in South Africa at the beginning of the 1980s, when the "United States [was] South Africa's largest foreign market and its leading source of imports" (Beaubien 1982, 9).[7]

The lines of such complicity in the economics of the distribution of popular culture to South Africa will keep emerging as I begin my reading of concurrent whiteness through an analysis of a single scene in *CE3K*. As Arnold Shepperson and Keyan Tomaselli point out, film attendance had plummeted in South Africa after the introduction of television in 1976. But after the merger of two of the largest distribution houses, by 1979 (when *CE3K* was released in South Africa), "the cinema-going public had increased to levels greater than before 1976" (Shepperson and Tomaselli 2002, 65–66). It is probably fair to say that as the events of 1976 in South Africa intensified international critique and calls for sanctions, so too were white South African anxieties about cultural isolation deepened. A consequent dialectical fetishizing of cultural products from the global north could, in part, explain the surge in film going in the late 1970s.

We should further note that the late 1970s were the years of Hollywood's science fiction boom, with films like *Star Wars* (1978) and *Superman* (1978) generating "a modern movie phenomenon [that rewrote] the economics of the Hollywood blockbuster and...heralded the age of [...] the 'megabuck' movie" (Tomaselli 1988, 154).[8] The product was not only the film, but the merchandise that accompanied it. And the consumption of these products represented South Africa's inclusion in global flows and consumption of culture. Keyan Tomaselli discusses a pertinent example of the unique merchandising strategy for the film *Superman* (1978), where merchandise was sold in advance of the film's release. In this case, Warner Brothers sold local manufacturing licenses on the Superman brand to various South African manufacturers. The products included "Superman play suits, pajamas, T-shirts, sheets, duvet covers, pillowcases, transfers, mugs, cups, plates, dolls, toy guns, bubble bath, soap, hair shampoo, toothpaste, bumper stickers and so on. [...] Local manufacturers found consumer response so great that they were able to export Superman merchandise to countries like the United Kingdom, Australia and the Philippines" (Tomaselli 1988, 155). While the South African consumer purchased this merchandise as a confirmation of his or her participation in the economies of a distinctly anti-communist global whiteness, the international consumer of South African products, even if unwittingly, supported apartheid's economy.

This is the context in which *CE3K* arrived in South Africa, itself already embedded in the fetishized commodity culture that this sci-fi era spawned.[9] Indeed, if we recall the profound disappointment that South Africans felt at their exclusion from the moon-landing, we can also speculate that fictional space worlds spoke directly to a sense of global participation in space technology. Furthermore, that the film was directed by Spielberg, whose *Jaws* had enjoyed massive success in South Africa, and that it was scored by John Williams (and 1979 was the year that Dolby Stereo sound was released worldwide), all added to the film's reputation as an exemplar of the most advanced technologies in the industry. Because the film was released with a two-year time lag in South Africa, it arrived with all these credentials—including two Oscar awards—already in place.[10]

Furthermore, to a white South African audience, this popular-cultural phenomenon was a projected alternate reality, one which allowed a radical dissociation from the contemporary realities of apartheid. This flight from reality cannot be seen in neutral political terms. The hyper-technology promised by films of the sci-fi boom resonated with white South African discourses of modernity that posited Africa as existing in a time before modernity. Technology was the overdetermined signifier of 1980s modernity: from the Walkman to the computer-game console, and what these mediated for white South Africa was culture from *elsewhere*. Add to that the aesthetics of science fiction and the movie theatre—a womb of sound and spectacle—becomes utterly dissociated from the concurrent historical circumstances existing outside its technological membrane.

To consider what it meant, then, for white South Africans in 1979 to watch this film, as an enactment of their participation in global media flows, I turn to Mieke Bal's notion that the site of political analysis of a work of art is the moment of "sentient engagement," which is to say, art is "empty as long as the act of viewing is not inherent to it, and that act is called upon to do political work" (2007, 23). If we use this idea as our starting point, asking what the sentient encounter of the white South African viewer with the film *CE3K* might have been, we enable a complex reading of the affective significance of participating in a global cultural space that imagined cultural life beyond the borders of the apartheid state. This might have included transformative moments, as discussed above, but the popularity of these films suggests that the affective experience had very little of the discomfort that the challenge of ideological transformation would surely include. Instead, the ease with which these "sentient encounters" occurred must speak to a harmony between white South African viewers and these American blockbuster films. That is, these films did not only appeal to the urge to flee the political realities of life in South Africa but also upheld the fantasy of a white suburban ideal that, I argue, was the banal substrate of apartheid life.

In this reading, I am not interested in the intentions and contexts of the film's production so much as in the limits those meanings collide with in the context of the film's screening in apartheid South Africa. That is to say, one would be hard-pressed to show that the film is in any way overtly complicit in apartheid ideology. Indeed, Spielberg himself is and was clearly no supporter of apartheid. In 1987, he co-signed a letter sent to Ronald Reagan requesting a cultural boycott of South Africa.[11] Furthermore, *CE3K* is notable for its conscious resistance to prior science fiction tropes of imperial conquest. In this film, it is the aliens that seek out humans, rather than the bold travel to the frontier that we recognize in films and television programs like *Star Trek* from this era. The aliens also turn out, here, to be peaceful beings whose intelligence and technologies far surpass those of their human counterparts. The film operates against the grain of both colonial metaphors and, one might argue, the ideal of the white middle-class American family. The male adult protagonist of the film, Roy Neary (played by Richard Dreyfuss), is not happy in his family circumstances and ultimately leaves this safe, domestic sphere for an adventure into outer space, while the lead female protagonist is a single, divorced mother, Jillian Guiler (played by Melinda Dillon).

More importantly to my discussion here, the film is not racist in its depiction of Black characters; on the contrary (and partly justifying my choice of it), it could even be said to provide a transformative representation in a scene in which an African American actor plays the role of an air traffic controller, something that would have been unthinkable in apartheid South Africa. Yet, this scene loses significance when we consider the fact that it is one of only two scenes representing African Americans in the film. The second scene representing African Americans in *CE3K* is the one that I wish to concentrate on here, keeping in mind what sort of "sentient engagement" a South African audience may have had with it.

In this scene, the protagonist, Roy Neary, obsessed with a mysterious hill that, he will later discover, is to be the site of the humans' close encounter with the aliens, starts to build a replica of the mound in his home. Desperate for materials to build with, Roy goes outside into the entirely white (if we judge from the neighbors presented in the scene), newly built suburb and starts throwing dirt, garbage, the neighbor's goose fence, anything he can get his hands on, into the house. That this is a newly built suburb highlights the suburban, capitalist promises of the era: the neighbors have a boat in their driveway; the family's television is on, drawing our attention to it; the children all have bicycles; the Neary family car is parked directly outside the house. In his desperate attempt to find materials for his structure, Roy Neary breaks all the codes of suburban life, and his wife's response shows this: rather than showing concern about what is ostensibly a total psychological breakdown, she is embarrassed by his behavior (and the scene ends with her packing her children into the car and driving off, never to be seen in the film again).

If we zoom in on one particular aspect of this scene, we notice something that might go unseen altogether: at one point, Roy struggles with a garbage collector, pulling a full garbage can out of the man's hands so that he can use the contents in his replica. The man gives up the struggle, lets Roy empty the garbage, shrugs, and continues towards the neighbor's house. It is a very short moment in a sequence of far greater significance in terms of the protagonist's psychological development, but one that warrants reflection. Both of the garbage collectors are Black. This might not be worth commenting on except for the fact that, as I mentioned above, other than the traffic control scene, these are the only Black characters represented in the film. The scene, a tableau of aspirant suburban, modern life in an entirely white neighborhood, with the detritus of that life being managed by Black men in tattered, dirty, overalls, could have been taken directly out of white suburban life in South Africa in the 1970s. The scene would glide smoothly and without disjuncture into the racial imaginary of South African audiences. The sentient engagement, in Mieke Bal's sense, of white audiences with a scene like this one (and these of course pervaded American film and television at the time), was one of implicit confirmation of their lives, lives lived at a remove from South Africa's Black majority who lived only under the conditions of a "bare life." The scene is only remarkable insofar as it would have gone entirely without notice in South Africa's racial imaginary at the time, as with so much else in the racial assumptions of popular culture coming into the country.

To be sure, this is just one scene in one film. Yet, part of the challenge of interpreting the political valence of whiteness is, as whiteness studies has often observed, isolating scenes for analysis in the thick, persistent, and pervasive stream of white dominance and privilege. This echoes what Lynn Spigel states of television depictions of space when she writes: "…more than just transmitting a privileged view of the universe, television offered the American public a particular mode of comprehension. It represented space, like everything else, as a place that the white middle-class family could claim as its own" (1991, 206). This scene, being set on earth, is even more overt in illustrating a ubiquitous assumption, in an American visual lexicon, that garbage collection is Black work. But more than that, those assumptions of race thicken as they flow into the prism of apartheid South Africa where the scene confirms the state's ideological apparatus that wishes to keep Black South Africans in bare life. The fact that the garbage collectors go almost unnoticed in a film dominated by white representation speaks to a structural absence of Black characters that would have easily slipped into South African white life. Concurrences would read the complicities and overlaps of South African whiteness with global racial and racist views and, as a method, would require us to rethink the comforting narrative of South African apartheid existing in a state of moral isolation. It would be a form of Derridean responsibility towards understanding these lines of complicity and would encourage media studies to complicate the narrative that global popular culture simply played a consciousness-raising role for white South Africans.

Concurrent Whiteness

To mine the full significance of what it might mean to bring the concept of concurrences to bear on dominant ideology, I wish to turn, briefly and by way of conclusion, to Hannah Arendt's analysis of the banality of evil in *Eichmann in Jerusalem*. Arendt notes Adolf Eichmann's seeming stupidity, his inability "to *think*, namely, to think from the standpoint of somebody else" (2006, 47). What emerges in the gap between thinking and thinking from the standpoint of the other is, of course, morality. For Arendt, Eichmann's "empty talk" (47) and tendency to express himself in "officialese" and clichés (46) evidences his incapacity for moral thought, a failure of moral imagination. The absence of moral imagination, paired with the reality that Eichmann was spared "the gruesome sights [, he] never actually attended a mass execution by shooting, he never actually watched the gassing process, or the selection of those for work" (87), means that Eichmann was able to sustain the immoral norm simply by not seeing, by not questioning, by not *thinking*. This is the banality of evil, to be sure, the total failure of imagining what is happening, concurrently, in parallel to one's life, the failure to understand one's moral responsibility for sustaining those parallel existences.

I do not want to draw facile comparisons between two very different, even if morally similar, systems as historically different as the Holocaust and apartheid. But Arendt's observations about banal evil are highly pertinent to my topic. This is not only because like so many German supporters of the Nazi party, who failed—or rather refused—to see/imagine/think the parallel lives of Jews, South African whites built their lives in the easy parallel spaces that the apartheid system had enabled. As such, they belonged to what Arendt calls a "new type of criminal, who is in actual fact *hostis generis humani*, [who thus] commits his crimes under the circumstances that make it well-nigh impossible for him to know or to feel that he is doing wrong" (274).

My aim has been to illustrate the complicity of concurrent whiteness that dominated the flows of popular culture during apartheid, not because it was overtly racist at the level of representation but because it so blandly and banally reproduced the structural absence of Black life, which at worst confirmed the normalcy of segregated white life under apartheid. To live a concurrent life, establishing through whatever means possible a wall between oneself and the other who suffers as a result of one's privilege, this is Arendt's "word-and-thought-defying *banality* of evil" (250, emphasis in the original). While nowadays the afterlife of apartheid is often only depicted in extreme registers, for example, in Anders Behring Breivik's reference to Afrikaner Nationalism in his notorious manifesto, or in the use of the term to describe the Israeli occupation in Palestine, these more extreme registers of apartheid's legacies miss the banal versions of whiteness that enjoyed, and often still enjoy, unfettered circulation, and that are in their banality all the more difficult to pinpoint and mobilize against. To relegate the history of South African apartheid only to the spaces of extremism is to miss the banal—and global—registers in which such moral failures occur.

Notes

1. While English- and Afrikaans-speaking whites had different processes of identification with "Europeanness" during this time (keep in mind that the moniker "European" was synonymous with "white" under apartheid), both can be seen to have invested in the idea of participating in global flows of technology and culture. The Afrikaans "political leadership took a cultural stance which glorified the technical achievements of the state's modernity" (Hyslop 2000, 37) and invested in the creation of an elite Afrikaans cultural scene, while English-speaking whites tended to identify more with the consumption of anglophone popular cultural flows. In both cases, though, what constituted white life was calibrated according to a cultural template that was European or Anglo-American, a cultural projection that racialized South African perceptions of consumerist modernity.
2. For example, even many American television sitcoms, which were not implicated by the Equity Ban, were dubbed into Afrikaans, with the original soundtrack running as a simulcast on radio.

3. Chris Hani had been the chief of staff to *Umkhonto we Sizwe* ("The Spear of the Nation"), the armed force of the African National Congress during apartheid that was, in those times, considered a terrorist group by the authorities.
4. I have already mentioned the *Moomin* animated series, but numerous Japanese animated versions of European stories, such as *Heidi, Nils Holgersson's Wonderful Journey through Sweden,* and others, were dubbed into Afrikaans. Van Staden makes the insightful point that "the fact that *Heidi* depicted Europe, but also lent itself to easy dubbing, enabled it to be hybridised with local politico-cultural concerns, particularly the apartheid government's obsession with European roots" (Van Staden 2014, 4).
5. The same amendment "cancelled the racial basis of censorship but retained the right to restrict films 'to persons in a specific category…or at a specific place.'" (Tomaselli 1988, 24), highlighting the state's anxiety as to which audiences (still racially imagined) could view what.
6. Given the ban on images of Nelson Mandela in South Africa at the time, as well as the censorship of anti-apartheid content, Theo Huxtable's "Free Nelson Mandela" poster became an intensely symbolic site of the stakes of cultural representation during apartheid.
7. "According to 1980 figures released by Pretoria's Department of Customs and Excise, the U.S. imported some $3.3 billion in metals and minerals such as platinum, diamonds, manganese and Kruger rand gold coins. In return, the U.S. exports to South Africa were valued at $2.5 billion for 1980, 17% higher than the previous year. U.S. corporate investment in South Africa has been growing at a rate of 25 percent per year, the most rapid rate of any U.S. foreign investment. U.S. capital earns an average rate of profit of 14.9% after taxes" (Beaubien 1982, 9).
8. See Tomaselli's discussion of the merchandise phenomenon and its economic implications in South Africa at this time (1988, 154–155). It is also worth noting that *CE3K* had a much larger budget (estimated at $19,400,870) than both *Star Wars* (estimated at $11,000,000) and *E.T.* (estimated at $10,500,000) (All estimates are from the International Movie Database, available at www.imdb.com).
9. *CE3K's* promotional material's use of the UFO as itself a cultural phenomenon is worth discussion but is beyond the scope of this paper. What is of relevance, however, is the large body of work that has analyzed the ways in which the UFO as cultural object was shaped by Cold War discourse at the time. See, for examples, McAllister and Eghigian 2022.
10. The film won Oscars for cinematography and sound effects (1978). It received nine Oscar nominations. As Robert Siegel points out, the film "was recognized as a motion picture phenomenon. From its first openings, critics were awed. Audiences responded in record numbers. There was vital

renewed interest in UFO phenomena, and even that phrase, *Close Encounters of the Third Kind* became part of our language" (2012).

11. See Clarke Taylor's "Film Directors Push Reagan for South Africa boycott," *Los Angeles Times*, November 3 1987. While Spielberg has been critiqued for appropriating histories and stories, in films like *Amistad* and *The Color Purple*, that belong within an African American repertoire, (Mcmullen and Solomon 1994), others have applauded the fact that these films offer important "civic education" for American audiences (Blum 1999).

References

Arendt, Hannah. 2006. *Eichmann in Jerusalem: A Report on the Banality of Evil.* New York: Penguin Classics.

Bal, Mieke. 2007. Lost in Space, Lost in the Library. In *Essays in Migratory Aesthetics: Cultural Practices Between Migration and Art-Making*, ed. Sam Durrant and Catherine M. Lord, 23–36. Amsterdam: Rodopi.

Beaubien, Michael. 1982. The Cultural Boycott of South Africa. *Africa Today* 29 (4): 5–16.

Bevan, Carin. 2008. *Putting up Screens: A History of Television in South Africa, 1929–1976.* Pretoria: University of Pretoria, Magister Dissertation.

Blum, Lawrence. 1999. Race, Community and Moral Education: Kohlberg and Spielberg as Civic Educators. *Journal of Moral Education* 28 (2): 125–143.

Bratton, Camille A. 1977. A Matter of Record: The History of the United States Voting Pattern in the United Nations (1946-1976). *Freedomways* 17 (3): 155–163.

Derrida, Jacques. 1994. *Specters of Marx: The State of Debt, the Work of Mourning, and the New International.* Translated by Peggy Kamuf. London: Routledge.

Evans, Martha. 2014. *Broadcasting the End of Apartheid: Live Television and the Birth of a New South Africa.* London: I.B. Taurus & Co.

Fur, Gunlög. 2017. Concurrences as a Methodology for Discerning Concurrent Histories. In *Concurrent Imaginaries, Postcolonial Worlds: Toward Revised Histories*, ed. Diana Brydon, Peter Forsgren, and Gunlög Fur, 33–57. Leiden: Brill.

Hyslop, Jonathan. 2000. Why did Apartheid's Supporters Capitulate? "Whiteness," Class and Consumption in Urban South Africa, 1985–1995. *Society in Transition* 30 (1): 36–44.

Krabill, Ron. 2010. *Starring Mandela and Cosby: Media and the End(s) of Apartheid.* Chicago: University of Chicago Press.

McAllister, Matthew P., and Greg Eghigian. 2022. Flying Saucers and UFOs in US Advertising During the Cold War, 1947–1989. *Advertising & Society Quarterly* 23 (3). https://doi.org/10.1353/asr.2022.0028.

Mcmullen, Wayne J., and Martha Solomon. 1994. The Politics of Adaptation: Steven Spielberg's Appropriation of *The Color Purple*. *Text and Performance Quarterly* 14 (2): 158–174.

Nixon, Rob. 1994. *Homelands, Harlem, and Hollywood: South African Culture and the World Beyond*. London: Routledge.

Popescu, Monica. 2007. Waiting for the Russians: Coetzee's *The Master of Petersburg* and the Logic of Late Postcolonialism. *Current Writing: Text and Reception in Southern Africa* 19 (1): 1–20.

Reed, Gail Ann. 1977. *Southern Africa: The United States Record at the United Nations, 1976*. The Africa Fund/American Committee on Africa.

Shepperson, Arnold, and Keyan G. Tomaselli. 2002. Restructuring the Industry: South African Cinema Beyond Apartheid. *South African Theatre Journal* 16 (1): 63–79.

Siegel, Robert. 2012. Making *Close Encounters of the Third Kind*. Accessed June 23, 2022. http://www.blu-ray.com/news/?id=7977.

Spielberg, Steven (dir.) 1977. *Close Encounters of the Third Kind*. Columbia Pictures.

Spigel, Lynn. 1991. From Domestic Space to Outer Space: The 1960s Fantastic Family Sit-Com. In *Close Encounters: Film, Feminism, and Science Fiction*, ed. Constance Penley et al., 205–236. Minneapolis: University of Minnesota.

Tomaselli, Keyan. 1988. *The Cinema of Apartheid: Race and Class in South African Film*. Chicago: Lake View Press.

Van Staden, Cobus. 2014. Moomin/Mūmin/Moemin: Apartheid-era Dubbing and Japanese Animation. *Critical Arts* 28 (1): 1–18.

CHAPTER 4

Settler Colonial Solutions to Settler Colonial Problems: Settler Cinemas and the Crisis of Colonization of Outer Space

Janne Lahti

Settler colonialism is an ongoing reality in many places around the world, and accordingly, scholars have tended to emphasize its permanence as a structure shaping people's lives, framing histories, and determining power relations in societies (Wolfe 2001, 2006; Kauanui 2016; Tuck and Wang 2012; Veracini 2010, 2015; Lahti 2017a). Yet, failure and crisis are also integral part of settler colonization as a historical process, as are environmental crises such as the Dust Bowl in North America. Settler colonialism tries to master the land, use it, and derive bounty from it, putting tremendous strain on and wrecking environments. This chapter examines settler colonial crisis, failures, and responses by reading speculative fiction. It looks at how the recent blockbuster films *Ad Astra* and *Interstellar* tackle imaginary settler projects in outer space in the context of settler failures, environmental problems, and the precariousness of habitability. These films deal in varied ways with far-migration, settler quests for land and living space, and settler understandings and relations to the land. In short,

J. Lahti (✉)
University of Helsinki, Helsinki, Finland
e-mail: janne.lahti@helsinki.fi

J. L. Hennessey (ed.), *History and Speculative Fiction*,
https://doi.org/10.1007/978-3-031-42235-5_4

typical settler colonial scenarios. They also discuss settler mindsets, instabilities, and the destruction of worlds, and their solutions and consequences. Put succinctly, they deal with settler colonial solutions to settler colonial problems, or, in other words, answers to settler colonial failures or potential failures. Thus, they address the precariousness of settler projects and how settler colonialism handles crises, how it reacts to problems of its own making, and what kind of solutions it envisions.

In *Interstellar* (dir. Christopher Nolan, 2014), the Earth is dying, the soil is exhausted, and the settlers are trying to cope with the calamity.[1] First, they try to change themselves, work collectively in conserving what remains and modifying their lifestyles to meet the worsening climate conditions. But they are failing, or, in another perspective, the planet is failing them. The crisis is seen as the fault of the planet no longer supporting settler lifestyles. Settler colonialism here is in denial that it is actually the cause of this crisis. In many ways, *Interstellar* addresses our most urgent current political question, global warming and the environmental crisis. But the solution it actually offers is more settler colonization. Environmental crisis is just another obstacle for the settlers to overcome without questioning their actions as settlers. In the process, the film dismisses the option of adjusting to the worsening conditions that the settlers' actions have caused. This kind of adjustment is useless. Regenerative practices or possible adaptations are deemed as failures. In fact, in the process of trying such measures, the settlers have instead lost their true selves. They have even denied their own heritage as pioneers; they shy away from their roots, they have canceled their own histories (censoring pioneer narratives in schools), and they have stopped searching for new lands. In a sense, they have stopped being settlers. Still, the Earth is finished; it can no longer support settler societies and merely offers misery and death in turn. So, the settlers need to rediscover their pioneer spirit and go to outer space to search for new lands to settle. And this is what takes place during the course of the film, a depiction of a dangerous, and ultimately successful, search to rediscover one's settler spirit and roots, and in the process one's settler future. All along, nature is an obstacle to overcome and to tame and control.

In *Ad Astra* (dir. James Gray, 2019), there is also a crisis, but its premises are different. The Earth is not failing, but the settlers seem to be. But not on Earth. Rather, their quest to search for new frontiers is threatening all life in the entire solar system. Furthermore, failure here is both personal, not collective, and ambivalent. The plot revolves around the mysterious Lima Project, a settler venture to search for lifeforms and to study the potential for settlement in distant corners of our solar system. Something has gone wrong with the project, and contact with it has been lost. A generation later, the fate of the Lima Project remains unknown, while mysterious power surges crossing space suddenly threaten all human life. Settler existence on Earth, the Moon, and Mars is in danger. It seems as if the environment in outer space is becoming uninhabitable, that the solar system is repelling human settlement, and that a crisis of epic proportions is imminent. Yet, it becomes apparent that settler actions—in this case, individual actions, not settling as a collective endeavor—have spurred this crisis. The crisis is connected with the Lima mission.

In *Ad Astra*, settlers have already advanced to the Moon and to Mars and are inhabiting both places. Even Mars has adults who were born there. And as settlers, they are on the move, looking for more potential real estate to conquer and possible other intelligent life forms—and potential rivals—on these outer frontiers. While the whole settler world is facing a crisis, settler colonization as such is never in doubt—meaning there is no option of regenerative practices, or turning back, of retracting one's steps in the face of peril. Settlers are and remain settlers in search of new frontiers. They don't need to adjust, even in the face of a crisis. But the crisis still needs to be averted. And for that, one mad settler hero needs to be reined in. The narrative here is told as a descent into the heart of darkness as our protagonist goes in search of his estranged father, a settler hero who led the Lima Project, to the outskirts of Neptune. During the journey, he faces his own demons, confronts his father—a settler hero gone mad—and traverses a settler outer space that is mired with conflict, where people seem miserable and disillusioned. The settler space here is a forlorn space.

In both *Interstellar* and *Ad Astra*, the remedy for settler colonial crisis seems to be more settler colonialism: just keep on colonizing. This involves a denial of the fundamental problems settler colonization causes. Crisis is never seen as the fault of settler colonialism; there is nothing fundamentally/structurally at fault with the settler premise of taking and using more land. Instead, it is the land or an unstable individual that is the cause of problems. Calls for ever more expanding frontiers to colonize are the solution when settler ambitions clash with uncontrollable realities. These films show how settler projects refuse to end, to die out, or to turn back, and how settler colonialism rejects critical self-reflection in the face of human and environmental disasters.

Settler Colonial Trajectories in History and Film

Historically, settler colonialism refers to a distinctive form of colonialism where the settlers aim to replace the Natives/previous residents and capture terrestrial and maritime spaces with the intention of making them their own. It involves conquest, long-range migration, permanent settlement (or at least the intent of such), elimination of Natives, and the reproduction of one's own society on what used to be other people's lands. According to Patrick Wolfe's classic definition, settler colonialism differs from other forms of colonialism in that it is not primarily an effort to build a master-servant relationship interested in exploitation of Native labor or the extraction of natural resources but instead is more concerned about replacement and access to territory, the land itself. Wolfe underlined that as "settlers come to stay: invasion is a structure not an event" or a series of isolated events. It "destroys to replace," introduces "a zero-sum contest over land," and is characterized by a "logic of elimination," a sustained institutional tendency to eliminate the Natives who stand in the way of settlers' ambitions of making the land their own (Wolfe 2001, 868; Wolfe 2006, 388).

Settler colonialism has been integral to the making of a competitive, integrated, and interlinked global order in the modern period. In short, settler colonization stormed, altered, and remade much of the world, trampling Indigenous lands and turning them into settler spaces, and replacing Indigenous peoples with settlers. Settlers imposed their rule and seized land across North America, from Alaska to the Canadian Plains all the way to Newfoundland, and from California to Florida, the Pacific Northwest to New England. They did so also in Australia, New Zealand, southern Africa, and French Algeria, as well as in northern Fennoscandia. Other settler projects remade Latin America as its states sought to whiten their societies by bringing in more European settlers (Castellanos 2017; Taylor and Lublin 2021). Russian state-led settler colonization efforts remade the Caucasus and the Siberia, while Japanese settler colonialism penetrated Korea and Manchuria. Germans also initiated settler projects in the German-Polish borderlands and in Southwest Africa (Belich 2011; Lahti 2019; Fujikane and Okamura 2008; Uchida 2014; Lu 2020; Hennessey 2020; Blackler 2022; Cavanagh and Veracini 2017; Barclay 2018a).

Many of these settler invasions, and many others, acted as "settler revolutions," typically characterized by explosive colonization, supercharged, exponential growth stemming from a synergy of ideological and technological shifts and following a cycle of expansion and integration (Belich 2011, 9). At the local level, the settler revolutions acted and felt more like shockwaves. They were networked, entangled, and co-produced mechanisms that altered the face of our planet. They carried tremendous human and environmental consequences, subjecting the land to excessive use, draining its resources and altering climates.

Settler colonialism also created global settler colonial cinemas with interlinked themes and joint narratives (Lahti and Weaver-Hightower 2020; Limbrick 2010; Columpar 2010; Veracini 2011). As a structure, settler colonialism needs to be continually reasserted and legitimized, as the settler's position on the land remains constantly challenged. The settler story must be recreated and retold because settlement is never fully accomplished. It is in this context that film as a genre and an industry developed, ascended, and morphed into the digital medium of the twenty-first century. Ever since their beginning, movies have captured and furthered the global settlement project, spreading its goals and normalizing its modes of domination when providing viewers with a recreation of the historic events of settlement (Weaver-Hightower and Lahti 2020, 3). Collectively, these films have juxtaposed and brought into contact different kinds of settlers with different kinds of experiences in different spaces. And in recent years scholars have interpreted these rich and diverse facets of settler cinemas, focusing on form, representation, reception, and production. Some have addressed urban settler environments, as in *Pépé le Moko* (dir. Julien Duvivier, 1937) set in French Algeria or in modern Taiwan in films such as *Home Sweet Home* (Jia zai Taibei, dir. Bai Jingrui, 1970) and *The Land of the Brave* (Long de chuanren, dir. Lee Hsing, 1981) (Flood 2020; Tsai 2020).

Arguably, the prototypical formula of settler cinema still remains the classic Western with the open spaces and its setting of Indigenous-settler encounters in those vast landscapes. The Western's key elements follow a logic of destruction, substitution, and rebirth, a settler colonial narrative built on the mutually constitutive process of elimination of Indigenous peoples and the setting up of settler families, enterprises, and communities on "virgin soil." The Natives are an obstacle the settlers need to ride over. They are incompatible with settler spaces; they need to make room, disappear, by force if necessary, or assimilate, change their ways. The settlers need to arrive at a particular destination, claim, occupy, and "settle" it to bring it under their "civilization." Many classic Westerns repeat this formula or depict some aspects of it, none more so than the anthology *How the West Was Won* (dir. John Ford, Henry Hathaway, and George Sherman, 1962), showcasing a multigenerational saga of settlers sweeping the continent, reinventing the land, and giving meaning and identity to its peoples. In these classic depictions—including *The Searches* (dir. John Ford, 1956), *Bend of the River* (dir. Anthony Mann, 1952), *Red River* (dir. Howard Hawks, 1946), or *Wagon Master* (dir. John Ford, 1950)—the settlers struggle, but ultimately overcome obstacles. They tame the land, which is bountiful but also hard and demanding. Settlers build a connection to the land through their sweat and toil, and oftentimes through their blood. While the meeting of the settler and the wilderness creates a rough and violent temporary frontier, an outskirt of civilization, in the end, the settlers transplant their values, norms, and societies with them, and they and the land itself are remade and regenerated as a result of the settler colonial process (Lahti 2017b). It is these depictions of the march of civilization, the never satisfied appetite for new lands, that the Western has come to symbolize, while some Westerns also criticize the outcomes of this settler colonization process (Lahti 2020). Here, failure is not really an option. In a typical Western, there is no environmental crisis that would make the settlers question colonization.

Failure and Crisis in Settler Colonial Framing

But what about cinema and settler colonial failures? What types of failures are there in settler cinemas, and how are they addressed in films? As a narrative, settler colonialism rejects turning back or giving up. A typical colonial narrative, historian Lorenzo Veracini writes, has a "circular form" with a clear return, while a settler colonial narrative differs from this as it moves forward without a return (Veracini 2010, 96–98). Settler cinema is grounded on notions of forward advancement. Settlers travel, arrive, and carry their lifestyles, values, and sovereignty with them. The key ingredient of this kind of narrative structure is the settlers' quest to indigenize themselves. They both transform the environment by civilizing it, by making the place their own, and renew themselves to suit the land. They make claims that the land and the sacrifice and work they have put into this land have made them belong. This makes it hard for settlers to contemplate or confront failure. Settlers instead often commemorate select epic moments or struggles that work to establish their connection with the land. Often these moments relate to violence. As Mahmood Mamdani asserts, settlers are made by conquest, not by immigration. They have arrived to replace the previous residents and the settlers remain settlers through continuous privilege and difference between them and the natives. Both initial replacement and ongoing dominance require and rest on violence, physical and discursive. Settler presence is meant to be for good (Mamdani 2015). Central here is the idea of permanence and belonging to the land. It is the settler who is at home and who makes the future. Until, that is, the settler is threatened with being kicked out, in real life or on the silver screen.

While settler rule remains firmly entrenched in North America and Australia, for example, it has, officially at least, been overturned in Africa. But what exactly happens when settler projects end, when settlers face failure, and how has this in turn resonated in settler cinemas? For one thing, there are different kinds of failures, stemming from political changes or environmental crises. Broadly speaking, there is the scenario of settlers being ousted by rival settler groups replacing the previous ones, or by decolonization, the liberation struggle of the previously subordinated Natives reclaiming their space. And there is also the scenario of the land becoming uninhabitable, as happened in the Dust Bowl in North America in the early 1900s (Holleman 2018; Worster 1979; Wunder et al. 2001). This led to settler exodus, settlers leaving the land, but not the repression of the settler colonial narrative. Elsewhere, decolonization would become a fresh start, with the settlers and their regimes now being replaced in turn. This was the case in French Algeria, where settler departure created lingering bitterness, longing, and nostalgia. The settler narrative refused to die, but much of its resonance was found in France, the destination of the evicted settler diaspora, not in the former settler colony itself (Hubbard 2015; Barclay 2018b). Much the same happened in Finnish Petsamo, where the period of Finnish settler colonization proved short-lived from the 1920s to the early 1940s. When Finland lost this land to the Soviet Union in 1944, all Finnish settlers needed to leave. Nevertheless, the memory of Finnish settler colonization lived on in Finland. Diaspora was perceived among the former settlers as an injustice, a forceful capture of the settlers' rightful homelands (Lahti 2021; Kullaa et al. 2022; Stadius 2021).

It is settler mobility that enables substitution of the Natives, initiates a new era, and constitutes a rite of passage necessary to legitimate the settler claim to the land through blood, tears, perseverance, and triumph. It is mobility that also is key to understanding the ways settler cinemas grapple, approach, and cope with failure, including environmental crisis. While traditional Westerns seldom deal with giving up land as such, or of settler land dying, ghost towns form a key signifier of failure in Westerns, in films such as *Yellow Sky* (dir. William A. Wellman, 1948) or *The Law and Jake Wade* (dir. John Sturges, 1958). They are, after all, sites where settlement has ended, and from where settlers had had to leave. But these failures are not typically perceived as settler failures. Rather they stand for the land failing the settlers. The settlers had taken the land's bounty, usually minerals, until the land has run out. And then the settlers had gone elsewhere to do the same. Exhaustion of resources is seldom viewed as major existential crisis that would lead to questioning and rejecting the whole settler colonial premise. Ghost towns are rather viewed as testaments to settler energy and virility, showcasing the efficiency and skill that settlers can harness and use nature's bounty. Even when the land has failed the settler farmers, as in *The Grapes of Wrath* (dir. John Ford, 1940), and the individual settler is forced to leave his/her home, it is not the settler's fault, and there is little reason for self-critique. Rather than consider radical reorganization of settler lifestyles when facing the Dust Bowl, collective relocation remains the solution in this film. Facing crisis in film, settlers routinely relocate and find new Natives to replace and new lands to make their own—this is the foundational act that constitutes their volitional polities, as Veracini marks (Veracini 2020, 203). Flight somewhere else is thus hailed as a preferable solution to regenerative practices or possible adaptation to a changing environment, let alone acknowledging that the settler is the cause of environmental problems.

The Land Is Failing

Interstellar addresses a global environmental crisis as an existential crisis of a settler society. It focuses on the dire consequences of settler colonization as the land the settlers have appropriated is dying. But it refuses to see the crisis as settler-caused. Instead, it is the fault of the land. We witness this reality via a small community in heartland America forced to adapt. The protagonist, Cooper (played by Matthew McConaughey), is a farmer who hates farming. He is actually a prototypical frontier man: restless, individualistic, opinionated, and physical. He is also a former NASA pilot. The society he is forced to live in with his teenage son and daughter is very much focused on scaling down, on returning to the basics, on trying to understand the land and to survive on its terms. The need for higher education is reduced; there is no need for engineers, but instead the world needs farmers when it is running out of food. History also has been rewritten, and, for example, the Moon landings are taken out of schoolbooks as falsifications, explained as merely constituting propaganda against the Soviets. The collective good trumps the individualistic settler spirit. The societies on Earth form some kind of a collective organizational system, a world government, and there are no armies anymore. Yet the crisis is acute and relentless. Blight has taken wheat some years before, okra is also gone. There is only corn left, but it will soon also die. There seems to be no way out. It is clear that this austerity and adjustment is failing. Or it is clear to people like Cooper who still embody the old pioneer spirit. Cooper feels that "we have forgotten who we are. Explorers, pioneers, not caretakers." The system has stifled the pioneer spirit; it has denied all vision of expansion and grandeur, of innovation, and of settling.

There is a key scene early on in the film that shows how surviving environmental crisis through communal action, by scaling down, is not working. During a baseball game, a massive dust storm sweeps the community, and people evacuate in the wail of alarm sirens. Blowing winds take off the topsoil, causing suffocating dust to invade everywhere, including our protagonist's home. It looks very much like the beginning of the end for Earth as a wave of ever more violent storms loom on the horizon for humankind. Settlers need to abandon their home, the Earth, and find a new home. While the majority of settlers who have lost their way refuse to see this, Cooper does not. His beliefs get confirmed as Cooper stumbles on a secret NASA base, where he learns that dust storms and failing crops will soon render humanity incapable of surviving on Earth. But NASA has been proactive, working in secret. Since staying put is not an option, two plans have been drawn up. The first option concerns removing as many people from immediate danger to seek refuge in an orbiting space station (although how to actually do this, to send the ark into orbit is as of yet unknown, but by the movie's end the gravitational equation is resolved, and NASA's space ark helps save humanity from immediate danger). The second option, the Lazarus Project, revolves around colonizing another planet with selected human material—sending frozen embryos out, terraforming, and repopulating. NASA has already earlier launched twelve ships to scout twelve possibly suitable worlds in outer space for humanity's next home. Of these twelve exploratory manned missions, three have activated their beacons, meaning they have found something meriting further investigation. Cooper is lured to join this mission, to investigate these beacons. Abandoning his life on and family on Earth, he goes to find new pastures to settle.

As Veracini remarks, *Interstellar* outlines a markedly settler colonial solution to crisis—"if the land turns against you, appropriate another." Indeed, the proposed remedy is not regenerative practices or possible adaptation, a change of lifestyles and a fundamental reordering in the understanding of the land (Veracini 2020, 204). No, the settlers' relationship to the land is never questioned. Instead, the proposed solution is to abandon the dying planet and seek new ones to settle. As the leader of the Lazarus Project, Professor Brand (Michael Caine) puts it, "We're not meant to save the world, we're meant to leave it." With this solution, the settler is not expected to change his/her outlook on land or to even question settler identity and practices. Rather, in order to survive, the current generation needs to return to the pioneering settler spirit of the previous generations, who originally settled these lands on Earth and put them to use. The current generation needs to rediscover its true settler spirit.

Cooper sets out on an epic journey to save the future of humankind, in this case, settlerkind. Initially his quest in outer space seems to result in failure as he passes through a wormhole into another galaxy with a planetary system orbiting a supermassive black hole called Gargantua. The hopes and dreams of new settler colonies are crushed by the harsh realities of the environments he encounters. Following the beacons, Cooper and his crew first seek out new settler futures on a watery planet where giant tsunami waves regularly crush everything in their path. There is no land there, and thus no settler futures. They next land on an icy, snow-covered planet where they encounter the previous explorer NASA sent, Mann (Matt Damon). It turns out that Mann has gone mad, falsifying data to get NASA to rescue him. He orchestrates a violent episode, seeking to hijack Cooper's spaceship, but fails. Regardless, it has become obvious that this planet too is unsuited for settler life. There is no surface, no air to breathe, and no future.

Ultimately, Cooper discovers a solution and returns to the new city in the sky, the secret NASA ark that by now has been launched to space. It sustains human life while a suitable new planet for settlerkind is also found. During his search for a new settler home, Cooper has also rediscovered his true self. He is a true explorer and pioneer and cannot permanently remain in the NASA ark on what is left of the Earth. Rather, the new frontier and its freedom beckon him. He needs to be at the forefront of the settler project. Cooper ventures to the new planet and joins the first pioneers in its settler colonization. It is a tough environment, but it awaits the settlers to take over, to tame the land, and to make it their own. Much like the older generations once did with their wagon trains on Earth.

In the end, humanity is on its way to new settler destinies. But it might also be on its way to wrecking new planets, to exhausting new environments. Earth is done for; now other planets stand in front of the settlers' insatiable gaze. Surely they will be exhausted eventually too. But the remedy is also there, now firmly established, to go search for new planets to consume. Settler colonization is truly never-ending.

Into the Heart of Darkness

In *Ad Astra*, the Lima Base is a settler outpost where things have gone terribly wrong. Not only can the mysterious power surges crossing space that suddenly threaten all human life be traced to it; already earlier some of the explorers in the Lima mission wanted to turn back, resulting in violence and tragedy. Their leader, H. Clifford McBride (Tommy Lee Jones), the famous settler hero revered across the world, refers to this attempt to turn back as "mutiny," to which he had responded by killing the mutineers. At some later point, closer to the present, even those faithful to Clifford had turned against him, with equally devastating consequences. There simply is no room for turning back in settler colonization. All this information Clifford shares with his son, the protagonist, Roy McBride (Brad Pitt). As the two finally meet after many years (in the last third of the movie), Clifford is all alone at the Lima space station. He is a man torn by his violent actions and his own stubborn quest to colonize and explore. He is vacillating between feelings of failure and desperation and mania to push onward, to explore new frontiers in search of intelligent life. He claims he is on a "one-way voyage," a man on a mission, a man with a destiny. He blames the others in his original group of cowardice, as lacking the necessary motivation and determination to do the hard work that colonization requires. They had caused the failure by wanting to return, by not believing in their mission. It is evident that Clifford is actually a man made insane by his colonial mission.

Like Kurtz in Joseph Conrad's famous colonial novel *Heart of Darkness*, the outlandish greed, unrestrained dreams, catastrophic failures, and extreme horrors tied with the colonization have done Clifford in. He has no way to go, no redemption, and no return. He is a man utterly alone and broken by his own mania over colonization. Even meeting up with the son he abandoned years ago can only bring temporary solace. For a brief moment, Clifford and Roy seem to bond, to share a mindset, and to dream of discoveries. But Roy soon recognizes how his father is so lost in his own mind and obsessed with finding new lifeforms that he cannot appreciate what he has already mapped and collected data on, the seemingly empty places on the outer edges of our solar system. In his contemplation, Roy seems to imply that what his father had indeed found were empty spaces, terra nullius, more potential spaces for the settlers to claims as theirs, exploit resources from, and possibly even inhabit in some form or another. While this is merely implied, between the lines, the film seems to suggest that the lack of other intelligent life forms might actually be a good thing, as these unoccupied spaces are available for humans to take over.

If Clifford is Kurtz, the great colonial hero gone crazy, sick, and irredeemable, then Roy is Marlow, Conrad's alter ego. While Conrad worked as a riverboat captain in 1890 for a Belgian trading company in King Leopold II's Congo, his novel from 1899 depicts a journey to the soul of European colonization in Africa, to a cruel savage terrain where violence permeates interpersonal and intergroup relations. Marlow embarks on a mission and heads deep inland, upriver. He is to reach the station of the prodigy of the European trading company, the renowned yet mysterious Mr. Kurtz. Kurtz is referred to as a prophet-like figure with great intelligence and grand designs. However, Kurtz has been cut off from communications and feared to have fallen gravely ill. As Marlow reaches Kurtz, he finds a fanatical, dying visionary who has made himself a god-like figure among local tribes (by introducing manufactured weapons) and raided the countryside for ivory. Kurtz has created a death zone in the heart of Africa. Yet, Marlow still speaks of the dying Kurtz as a great man, a highly intelligent pioneer of sorts, albeit of a deranged and lethal kind whose soul had gone mad (Conrad 1899; Jasanoff 2017).

Conrad's work remains possibly the best-known work of fiction dealing with colonial violence. In *Ad Astra* there is a similar kind of journey, a descent into darkness, but this time in colonized outer space. Roy is sent to open contacts with his father, to message him from Mars. And that is all he is meant to do. The purpose of this secret mission, not revealed to Roy by Spacecom, the colonial space company, is to actually eliminate Clifford and to stop the mysterious power surges by sending in a strike force to nuke the Lima Base. But Roy actually ends up going to meet his father in person, going against his orders by capturing a space shuttle in Mars and venturing alone to confront Clifford. This happens when he starts to realize there is something wrong with Clifford and that he needs to sort it out personally. Before that Roy travels through a bleak and unforgiving colonial terrain, much like Marlow does. He himself is also lost when starting the mission. Roy is unemotional, disillusioned, and estranged from those closest to him on Earth. In short, Roy is an alienated, anxious man who has lost his purpose and will to live. But through his mission, he regains his life powers. He gains redemption by confronting his father, the great explorer and pioneer hero gone crazy, and by putting an end to the surges, putting the colonization project back on track from the brink of disaster.

When Roy lands at the Moon, the first stop on his journey, he enters a commercialized and commodified colonial space, catering to tourists and capital. While the official marketing slogan on the hangar promotes Moon as a place "Where the World Comes Together," in reality it is a scene of settler conflicts. The Moon is a war zone, a battleground between different settler factions, with rebels ominously dubbed by the establishment as "pirates." They are the Native Americans to the white settlers. The situation is even referred to as such: "It's like the Wild West out there," one Moon resident states, referring to spaces outside the main settlement. When Roy is transported to his outgoing space shuttle on the dark side of the moon and needs to cross "no man's land," his crew is attacked by these pirates. It is done in a very classic Indian attack style, copied from John Ford's *Stagecoach* (1939) and numerous other Westerns. Pirates charge after the moving space vehicles that Roy is being transported in, chasing them across barren lunar plains. Roy is saved by the "cavalry," in this case firing from base that repels the attackers. Like in many classic Westerns, no explanation to why these pirates act the way they do is given. What motivates them, what has caused the hostilities remains a mystery. There is simply competition over resources and land. Violence is in fact represented as an ordinary circumstance, as part of the settler colonial everyday, as part of the process of replacement where settlers eliminate their rivals and make the land their own. Clearly, this process is still unfinished on the Moon. Rather the Moon is rife with conflicts settler colonization has brought.

As Roy leaves the Moon behind him, the audience learns how humans are "world eaters," who explore, take, and consume all they can. Next up is Mars, the "last secure hub we have" as one colonizer calls it. In its underground settler society, some 1100 people live in a bleak, concrete world. There is no sun, air, or flowers, except in artificial simulations projected on walls inside the compound. Mars seems much like a wasteland, a last outpost on the settler frontier. And it is not a happy family settlement, or a thriving frontier town we have seen in so many old Westerns. It is barely hanging on, almost a ghost town already, although it has adult settlers who are native-born in Mars. People there seem scared, lonely, and lacking a sense of purpose. "Is it really worth saving?" is a question Roy never voices. Instead, he goes to save it and the whole settler world. As Clifford spirals to his demise in open space, Roy destroys the Lima project and stops the surges. He makes it back home to Earth to live a happier life than before. Like Cooper in *Interstellar*, he too has gained his true identity through his daring adventure in space in the service of settler futures.

Conclusion

At the end of these two films, settler worlds are saved, reinvigorated, and reenergized. Settler colonization is alive and well. Through their continual replaying and retelling of settlement stories films like *Interstellar* and *Ad Astra* keep on reminding all that settlement is never finished, but constantly ongoing. They have affirmed settler colonialism's refusal to die. *Ad Astra* and *Interstellar* provide a window to the settler mindset and identity, what is essential about settling. They speak to a common, international, audience, and use a shared "language" of settler colonialism in doing so: the stories of empty lands, settler civilizations and righteousness, and of forward advancement and exploration. *Ad Astra* and *Interstellar* comment on these settler colonial themes of land use and environmental crisis. They reaffirm and play with typical settler colonial narrative outcomes, the themes of settler arrival, quest for land and family, ongoing settlement, and displacement. While *Ad Astra* provides some critical commentary of settler actions as "world eaters," neither film seriously questions the settler colonial logic, let alone promotes the end of settler colonization. *Interstellar* openly advances the move for further settler colonization, endlessly, while *Ad Astra*, by saving settler space from doom, normalizes settler colonization as an imperfect but normative state of the human condition. Advancing forward is what the settler, and thus human society, is all about.

Note

1. My reading of *Interstellar* owes much to Lorenzo Veracini's excellent study of *Interstellar* and *Cowboys and Aliens* (2020).

References

Barclay, Fiona, ed. 2018a. Settler Colonialism and French Algeria. *Special Issue of Settler Colonial Studies* 8 (2): 115.

———. 2018b. Remembering Algeria: Melancholy, Depression and the Colonising of the Pieds-Noirs. *Settler Colonial Studies* 8 (2): 244–261.

Belich, James. 2011. *Replenishing the Earth: The Settler Revolution and the Rise of the Angloworld*. Oxford: Oxford University Press.

Blackler, Adam. 2022. *An Imperial Homeland: Forging German Identity in Southwest Africa*. Penn State University Press.

Castellanos, M. Bianet. 2017. Introduction: Settler Colonialism in Latin America. *American Quarterly* 69 (4): 777–781.

Cavanagh, Edward, and Lorenzo Veracini, eds. 2017. *The Routledge Handbook of the History of Settler Colonialism*. New York: Routledge.

Columpar, Corinn. 2010. *Unsettling Sights: The Fourth World on Film*. Carbondale: Southern Illinois University Press.

Conrad, Joseph. 1899 [2020]. *Heart of Darkness*. London: Alma Classics.

Flood, Maria. 2020. From Colonial Casbah to Casbah-Banlieue: Settlement and Space in *Pépé le Moko* (1937) and *La Haine* (1996). In *Cinematic Settlers: The Settler Colonial World in Film*, ed. Janne Lahti and Rebecca Weaver-Hightower, 177–188. London: Routledge.

Fujikane, Candace, and Jonathan Y. Okamura, eds. 2008. *Asian Settler Colonialism: From Local Governance to the Habits of Everyday Life in Hawai'i*. Honolulu: Hawai'i University Press.

Hennessey, John. 2020. A Colonial Trans-Pacific Partnership: William Smith Clark, David Pearce Penhallow and Japanese Settler Colonialism in Hokkaido. *Settler Colonial Studies* 10 (1): 54–73.

Holleman, Hannah. 2018. *Dust Bowls of Empire: Imperialism, Environmental Politics, and the Injustice of "Green" Capitalism*. New Haven: Yale University Press.

Hubbard, Amy L. 2015. *Remembering French Algeria: Pieds-Noir, Identity, and Exile*. Lincoln: University of Nebraska Press.

Jasanoff, Maya. 2017. *The Dawn Watch: Joseph Conrad in a Global World*. London: William Collins.

Kauanui, J. Kēhaulani. 2016. "A Structure, Not an Event": Settler Colonialism and Enduring Indigeneity. *Lateral* 5 (1).

Kullaa, Rinna, Janne Lahti, and Sami Lakomäki, eds. 2022. *Kolonialismi Suomen rajaseuduilla [Colonialism in Finland's Border Regions].* Helsinki: Gaudeamus.
Lahti, Janne. 2017a. What is Settler Colonialism and what it has to do with the American West? *Journal of the West* 56 (4): 8–12.
———. 2017b. Settler Passages: Mobility and Settler Colonial Narratives in Westerns. *Journal of the West* 56 (4): 67–76.
———. 2019. *The American West and the World: Transnational and Comparative Perspectives.* London: Routledge.
———. 2020. The Unbearable Settler West in *The Ballad of Buster Scruggs*. In *Cinematic Settlers: The Settler Colonial World in Film*, ed. Janne Lahti and Rebecca Weaver-Hightower, 77–86. London: Routledge.
———. 2021. Settler Colonial Eyes: Finnish Travel Writers and the Colonization of Petsamo. In *Finnish Colonial Encounters: From Anti-Imperialism to Cultural Colonialism and Complicity*, ed. Raita Merivirta, Leila Koivunen, and Timo Särkkä, 95–120. New York: Palgrave.
Lahti, Janne, and Rebecca Weaver-Hightower, eds. 2020. *Cinematic Settlers: The Settler Colonial World in Film.* London: Routledge.
Limbrick, Peter. 2010. *Making Settler Cinemas: Film and Colonial Encounters in the United States, Australia, and New Zealand.* New York: Palgrave.
Lu, Sidney Xu. 2020. *The Making of Japanese Settler Colonialism: Malthusianism and Trans-Pacific Migration, 1868–1961.* Cambridge: Cambridge University Press.
Mamdani, Mahmood. 2015. Settler Colonialism: Then and Now. *Critical Inquiry* 41 (3): 596–614.
Stadius, Peter. 2021. Petsamo 1920-1940: Rhetoric of Colonialism and Finnishness. *Journal of Finnish Studies* 1–2: 112–135.
Taylor, Lucy, and Geraldine Lublin. 2021. Settler Colonial Studies and Latin America. *Settler Colonial Studies* 11 (3): 259–270.
Tsai, Lin-chin. 2020. Negotiating Between Homelands: Settler Colonial Situation and Settler Ambivalence in Taiwan Cinema. In *Cinematic Settlers: The Settler Colonial World in Film*, ed. Janne Lahti and Rebecca Weaver-Hightower, 99–112. London: Routledge.
Tuck, Eve, and Wayne K. Wang. 2012. Decolonization is not a Metaphor. *Decolonization: Indigeneity, Education, and Society* 1 (1): 1–40.
Uchida, Jun. 2014. *Brokers of Empire: Japanese Settler Colonialism in Korea, 1876–1945.* Cambridge, MA: Harvard University Press.
Veracini, Lorenzo. 2010. *Settler Colonialism: A Theoretical Overview.* New York: Palgrave Macmillan.
———. 2011. District 9 and Avatar: Science Fiction and Settler Colonialism. *Journal of Intercultural Studies* 32 (4): 355–367.
———. 2015. *The Settler Colonial Present.* New York: Palgrave Macmillan.

———. 2020. Settler Evasions in *Interstellar* and *Cowboys and Aliens*: Thinking the End of the World is Still Easier than Thinking the End of Settler Colonialism. In *Cinematic Settlers: The Settler Colonial World in Film*, ed. Janne Lahti and Rebecca Weaver-Hightower, 203–214. London: Routledge.

Weaver-Hightower, Rebecca, and Janne Lahti. 2020. Introduction: Reel Settler Colonialism: Gazing, Reception, and Production of Global Settler Cinemas. In *Cinematic Settlers: The Settler Colonial World in Film*, ed. Janne Lahti and Rebecca Weaver-Hightower, 1–10. London: Routledge.

Wolfe, Patrick. 2001. Land, Labor, and Difference: Elementary Structures of Race. *American Historical Review* 106 (3): 866–905.

———. 2006. Settler Colonialism and the Elimination of the Native. *Journal of Genocide Research* 8 (4): 387–409.

Worster, Donald. 1979. *Dust Bowl: The Southern Plains in the 1930s*. Oxford: Oxford University Press.

Wunder, John R., Frances W. Kaye, and Vernon Carstensen, eds. 2001. *Americans View Their Dust Bowl Experience*. Boulder: University Press of Colorado.

CHAPTER 5

The Weirdness of White Strangers: Imaginations of Westerners in Southeast Asian Lore and Tradition

Hans Hägerdal

Introduction: Occidentalism in Asian Lore

Since the late 1970s, a lot of ink has been spilled over Orientalism in the sense of stylized Western images of the East. In the famous analysis of Edward Said, Orientalism becomes the fantasized abode of a set of cultural essences, such as despotism, lasciviousness, depravity, fanaticism, and immutability. The Orient, in this model, becomes the Other, an entity fundamentally different from the ostensibly rational and civilized West. All this links to an intellectual genealogy that is reified through the Western knowledge production. In essence, the Western views of the East are conditioned by a will to maintain authority over the Other and are therefore intimately tied to the colonial and neo-colonial projects (Said 1978). The analysis of Said and later postcolonial scholarship has in turn been criticized for being too generalizing and essentializing. Some of this critique is no doubt valid, though the Said thesis is still interesting for pointing out

H. Hägerdal (✉)
Linnaeus University, Växjö, Sweden
e-mail: hans.hagerdal@lnu.se

J. L. Hennessey (ed.), *History and Speculative Fiction*,
https://doi.org/10.1007/978-3-031-42235-5_5

the close relation between claims of power and intellectual knowledge (Hamidi 2013).

The silent partner of this is, consequently, Occidentalism, meaning stylized images of the West. There may be two sides to this. Western writers over the centuries, who construct essentializing images of non-Western cultures, necessarily juxtapose the features of Chinese, Indian, or Islamic civilization with their own norms of normality. This is done either explicitly or tacitly. Orientalism is therefore generally paired with an unsaid and positively laden Occidentalism (Carrier 1995, 2–4). However, there is another and more anthropological side to it. Peoples outside the Western world are no less keen to define their own cultural specifics than are Europeans, Americans, and Australians. While terms such as nation and culture may not always translate easily into non-Western languages, expressions of self-identity are omnipresent as social strategies. In such expressions, the awareness of other cultures, especially when somehow menacing or competing, will likely lead to a juxtaposition. Anthropological research indicates that awareness of assertive Western culture may reinforce certain indigenous customs in order to highlight the distinction (Carrier 1995, 6–8).

The question is, nevertheless, whether Occidentalism can function the way that Orientalism does according to the Said thesis. The crux of it is the unbalanced power relations that have supposedly favored Western hegemonic claims for hundreds of years. The global presence of Western-derived culture, values, technology, and institutions has certainly enticed non-Western cultures to launch a set of counter-discourses. Some of these are quite belligerent and essentialize the West as the perennial bogeyman (Buruma and Margalit 2004). But can this, in the end, parallel Western Orientalism? Despite the best efforts of Chinese, Arabic, and Indian leaders and intellectuals, a similar positional superiority vis-à-vis the West might be difficult to maintain—at least if we talk about the world that was transformed by imperialism in the nineteenth and twentieth centuries.

With all this in mind, it may still be worth the effort to analyze the features of Occidentalism in a historical region with long experience with Western influences. How were stylized images of Europeans constructed and juxtaposed to regional cultural values? The choice here is Southeast Asia, a region characterized by tropical and subtropical climate, early written cultures, the presence of important trading networks, and the influx of all world religions over the centuries. Europeans are known here since the early years of the sixteenth century, which suggests the question of how they are portrayed in works of art, literature, and history writing. For a brief chapter, the study can only be exploratory. My material is an array of chronicles and historical and legendary traditions, authored in the course of more than 400 years. While by no means exhaustive, the essay will exemplify the type of imaginations of Europeans in the eyes of Southeast Asian observers. In particular, I highlight the inclusion of Europeans in legendary and even fantastic contexts, overlapping what is termed "speculative fiction" elsewhere in this volume. This device will allow us to study the processes of stereotyping more clearly, as the authors are not restricted by exact historical accuracy and the images of the Western Other are rounded out using the author's imagination.

Europeans in Southeast Asia

Southeast Asia constitutes a subcontinent and an island world with many hundreds of languages and cultures. It is often defined in terms of external influences over the millennia, since it was an early cultural and commercial crossroads. Chinese culture imprinted parts of Vietnam, Indian Hindu culture pervaded most of mainland and part of insular Southeast Asia and later gave way to Buddhist influences, while much of the maritime world was converted to Islam. All these influences tended to be adapted to a local context that allowed for the perseverance of local beliefs and customs (Andaya and Andaya 2015, 53). While it is difficult to pinpoint a unanimous Southeast Asian identity, the mostly tropical and monsoon-dependent climate combined with the reliance on wet rice cultivation and the extraction of forest products in the main historical regions gave the area certain characteristics that set it apart from the South and East Asian cultures. Contacts between Southeast Asians and Europeans were almost nonexistent until the early sixteenth century. Nevertheless, Southeast Asia was among the very earliest regions in Asia to be subjected to colonial rule. The Portuguese established themselves in the Melaka Peninsula after 1511 and the Moluccas after c. 1522. The Spanish arrived in the Philippines in 1565 and the Dutch began to acquire outposts in the Indonesian Archipelago after 1605 (Andaya and Andaya 2015, 134, 159, 166; Tarling 1999, 10, 73, 184–5). Much of the early-modern colonial enterprise was directed at the securing of trade routes and monopolies, as the European seafarers were attracted by profits in spices, textiles, and later cash crops such as coffee. The large wave of colonial conquest, however, occurred during the nineteenth century when all of Southeast Asia with the exception of Siam fell under European or American rule. European ships and violent interventions were therefore a menacing reality for several hundred years.

The Europeans were intruders but also a source of power and wealth, and therefore potential allies or protectors. The history of what is today Indonesia, Malaysia, and Timor-Leste is filled with attempts by local groups to win the dangerous and potent foreigners to their side to fight local enemies. On the mainland, kings and pretenders used European mercenaries and adventurers to advance their interests. The well-known Burmese kings Tabinshwehti and Bayinnaung made good use of Portuguese soldiers in the sixteenth century (Harvey 1925, 155, 162). Spanish and Portuguese adventurers briefly set themselves up as kingmakers in Cambodia in the 1590s. French mercenaries helped establishing the Nguyen Dynasty in Vietnam around 1800. Europeans could also serve as valued councillors, such as the Greek-born Constance Phaulcon in Siam in the seventeenth century (Tarling 1999, 79, 101, 245). Last but not least, the commercial capabilities of European trading organizations made the white foreigners indispensable in many places.

General Ideas About the White Foreigners

Some trends in early Southeast Asian images of Europeans have been traced by Anthony Reid. As he notes, Southeast Asia had received faraway foreigners for over a thousand years when the Europeans arrived. Although their white skin might have evoked some curiosity, their physical features were not much different from people from West Asia—Persians and Arabs were at least semi-regular visitors in the Middle Ages. According to early reports of the encounter, the Portuguese were characterized as "White Bengalis" or "White Indians" and were likened to the fair-skinned seafarers from the Ryukyu Islands (Brown 1970, 151; Reid 1994). Large ships and artillery were no news for the Southeast Asians. What set them apart was rather the aggressive means they used to acquire bases and monopolies. Cannons were used much more efficiently than in the Southeast Asian kingdoms that they encountered, and the business-like agenda of the East India Companies was far more comprehensive than local commercial organizations.

The Westerners' general behavior is often commented upon. A famous but perhaps apocryphal story in the *Sejarah Melayu* (1612) tells how a Portuguese fleet officer in 1509 hung a golden chain around the neck of the chief minister of the Melaka Sultanate, thus provoking the anger of the bystanders for his rudeness, although hostilities were avoided for the time being (Reid 1994, 275). The Ambonese chronicle *Hikayat Tanah Hitu* (c. 1650) relates how the locals at first welcomed the Portuguese to their island in the Moluccas in the early sixteenth century, but that incident soon occurred: "It once happened that they got drunk and then went to the market to plunder and caused upheaval." The whites were therefore banished to another village where people were not Muslims and kept alcoholic drinks. "After some time there arose trouble, and they fought against them. They attacked each other, they won over each other, and it appeared as if the holy war had no end … The sound of their weapons can be likened to the thunder from heaven" (Manusama 1977, 92–3).

The royal chronicles of Cambodia contain a few episodes featuring Europeans. These chronicles were embellished over time, though ultimately referring to historical figures (at least after 1500). King Ponhea To had a short and turbulent reign (1628–31) as he was attacked by his uncle after an erotic indiscretion and forced to flee. A late version of the chronicle relates how the young fugitive hid on the top of a sugar palm and was tracked down by a party of European mercenaries. As the soldiers aimed with their muskets, the king pronounced with a loud voice:

Damned Europeans!
They pursue in order to shoot at the Lord
The august Raj Sambhar
Who has ascended to the midst of a sugar palm
Without fearing his might.
The Lord raises his hand
To make them hold back and wait (Mak Phoeun 1981, 391)

The protestations of the king were to no avail, as he was shot down by the Europeans after having delivered a moralistic poem. Although European mercenaries might have been present at the regicide, the scene of a king composing verses before his demise is, of course, fiction that emphasizes the dark aspects of employing white soldiers for hire. Although the story actually recounts an internal dynastic feud, the foreigners are depicted as a frightening tool to kill the legitimate ruler under the most ignoble circumstances.

Westerners are also sometimes inclined to break sexual norms. An undated Malay *syair* (song) known in several versions tells of Sinyor Kosta

(Senhor Costa), a young Portuguese merchant who arrives to an unnamed Southeast Asian port and sets up a trading factory. There he sets eye on Siti, a Sino-Burmese girl of outstanding beauty who is, however, married to a Cantonese man.

> Up to the alley he would walk
> To catch a glimpse of Siti in the upper room.
> His hat cocked and umbrella furled
> With sidelong glances he would look at her (Teeuw et al. 2004, 254)

Moreover, his passions are not restricted to Siti. He is also "seized by desire" when he once imbibes a large quantity of arrack and sees a pretty Balinese girl and "yearned to have some sports with her."

> The Balinese girl flung the cloth down,
> Saying, 'Don't fool around with me'.
> If your passion is aroused so strongly,
> Find someone else, don't pick on me. (Teeuw et al. 2004, 271)

The story emphasizes the uncontrolled behavior of the Portuguese, whose adultery with the Chinese Siti ultimately leads to misery. Though the author(s) of the song appears to feel sympathies with the intense passions of the wayward Westerner, the latter is also used as a literary device to embody impassionate and unbridled conduct and breach of sexual norms.

Another variant on the theme of wayward Westerners is found in the Burmese chronicle *Hmannan Yazawin* (1832). The Burmese ruler Tabinshwehti (1531–1550) gathered the major part of the country under his sway after centuries of political division. In this he was helped by Portuguese mercenaries. The chronicle alleges that the king, after a series of warlike exploits, began to consort with a young Feringhi (Portuguese) and fell from kingly virtue. The Portuguese rose to prominence due to his skills with the awe-inspiring firearms and taught his local wife to cook dishes from his homeland to serve the king, with great success. Eventually the foreigner prepared spirits that soon turned Tabinshwehti into an alcoholic who lost his wits. The royal entourage managed to exile the Portuguese in the end, but it was too late. Some of his chamberlains murdered the king, and the empire had to be constructed anew by his successor, Bayinnaung (1551–1581). There are some problems with the story, which does not entirely fit with contemporary sources (Harvey 1925, 160–1, 343). Anyway, the chronicler regards the Portuguese as harmful and norm-breaking figures with the potential for debauchery against Burmese people and creating harm for the country.

The Mythical Past

The image of the distant past is not merely a mosaic of mythologized persons and events but also a charter against which the present may be judged. In a Southeast Asian context, origin stories often play a seminal role in explaining the hierarchies and cultural specifics of a society. We may therefore ask if Europeans have a place in such stories. It should be remarked that Southeast Asian historiographies did not usually have the means to determine whether Europeans had been present since the sixteenth century or for much longer than that.

In most places, mythical origin stories make no references to Europeans. The chronicles of the mainland kingdoms typically start in a remote age and mix local folklore, Hindu myths, Buddhist legends, and the chroniclers' own constructions. Europeans come in late in the chronicles, and only sparsely, when dramatic and warlike events occur. The situation is different with historical and pseudo-historical narratives which emerged in the maritime world. Europeans occasionally occur in remote times, by implication long before the actual start of contact in 1509. Thus, the Malay epic, *Hikayat Hang Tuah,* lets its hero Hang Tuah act in a landscape that defies historical chronology in the Western sense. A loyal follower of the Sultan of Melaka, Hang Tuah travels to many countries, including India and China, where he meets hostile Portuguese seafarers whom he defeats through his bravery and use of magic spells. He also defeats a large fleet from Spanish Manila that attacks Melaka—long before the actual appearance of the Portuguese or Spanish in the region. Only after Hang Tuah's death are the whites able to invest Melaka through cunning means (Reid 1994, 282–3).

The *Hikayat Banjar* (1660s), which details the history of the largest kingdom of southern Borneo, takes for granted that Dutch people came to trade on the same premises as Chinese, Malays, Indians, and others, at a time that lies several generations before European contacts with Southeast Asia (Ras 1968, 263). Likewise, the Balinese legendary chronicle *Usana Jawa* (seventeenth century?) mentions a large number of ethnic groups associated with the Javanese Majapahit Empire (1293–c. 1520), which include the Hulanda people alongside Brunei, Bugis, Makassar, Tambora, and so on. The Hulanda are, of course, the Dutch who in fact only arrived in 1596; thus, there is no sense that the Dutch are newcomers to the islands. Curiously, the *Usana Jawa* asserts that the Hulanda were found to the east of Majapahit, perhaps since they were first established in the Spice Islands in the far eastern archipelago and only later founded Batavia (Jakarta) on Java (Warna 1986b, 93). Even more to the point are legends from the eastern parts of maritime Southeast Asia. Stories of old times told in East Timor sometimes mention a generic affinity between the Portuguese colonizers and Timorese peoples. The Mambai people see themselves as senior in a Timor-wide context, but at the same time, the power of the local elite is thought to emanate from the Portuguese, who are incorporated into a mythologized system. The Portuguese are no unqualified strangers to the land since they have parted from Timor in a remote time. Therefore, they are in fact younger sons returning to the land (Traube 1986, 52–4).

Overcoming White Strangers

As Anthony Reid has noticed, the idea of Westerners in early Southeast Asian texts tends to be either neutral or negative, with the negative images gaining ground over time. He finds almost no early text that puts European culture and ethnicity in a positive light (Reid 1994, 288). As we shall see later, this view may not be entirely correct, but it is apparent that many narratives construct an image of the white strangers as fundamentally aggressive and hostile. Given the centuries-long history of colonial and commercial encroachment, this may not be surprising. Against this background, there are many stories that tell of defensive stratagems against the invaders, stories that mostly have a historical background but may be less than accurate. Supernatural power is a recurring ingredient in the struggle; in many parts of Southeast Asia, personal prowess is associated with a concentration of cosmic power. Objects such as daggers, swords, and even bullets are supposedly imbued with magic power and represent a channel between the local/individual microcosm and the larger macrocosm (Andaya and Andaya 2015, 49–53). This is overlaid by religious antagonism: the Europeans are cast as the enemies of Islam or Buddhism, which had become dominant in Southeast Asia by the early-modern era.

The magic exploits of Hang Tuah against the Iberians have been mentioned. The part-legendary chronicle of Banjarmasin on Borneo, *Hikayat*

Banjar (1660s), tells of defiance against the Europeans in more subtle terms. The historical Dutch attack on Banjarmasin in 1636, a punitive expedition where four vessels shelled the royal city, is told in some detail, being the only substantial mention of Europeans in the chronicle. However, the calamitous nature of the attack is modified by the stance of the sultan, Marhum Panembahan. He solemnly forbids three of his chiefs to attack the ships, as he himself has the means to ruin them. Through his innate powers, he can make the Dutch lose their wits, and eventually, the attack is terminated and the ships withdraw. Though not actually denying the ruinous Dutch action, which causes him to search for a new capital, the kingdom is protected from worse devastation through macrocosmic intervention via the pious sultan (Ras 1968, 465).

A more ecology-oriented variant is told in Belu in Central Timor, which historically had to deal with both the Portuguese and the Dutch. Narratives recorded in the twentieth century tell of ancient invasions by Portuguese, who are commonly seen as dangerous and hostile figures in contrast to the trading Makassarese people from Sulawesi. The Belunese cannot match the Portuguese proficiency in firearms; instead, they know how to activate black bees and wasps through a peculiar ceremony. During the battle that follows, the insects swarm around the Europeans and force them to flee in haste (Spillett 1999, 143). Interestingly, a similar story is told about the Dutch war of conquest in the western part of the island in 1905–1906. The last ruler of the West Timorese inland realm of Sonba'i, Nai Sobe Sonba'i III, used the strength of the inherited ancestral knowledge to counter the invaders. When the whites, who are here likened to "white monkeys in trousers," advanced on the mountain stronghold Kauniki, the ruler directed the bees swarming in the eucalyptus groves against them. The Timorese defenders thus gained some time to regroup as the colonial soldiers fled head over heels, although the ruler was betrayed and captured in the end. Contemporary colonial reports do not mention any impact from bees, but the narrative provides a striking example of anti-colonialism intertwined with ancestral beliefs and forces of nature (Hägerdal 2017, 592; Fobia 1984, 102).

Bonding the Strangers

Such stories indicate a will to portray the Westerners as an essentially destructive force opposed to local norms. Though written or narrated at different times, the stories arose in a world where Europeans could not be easily expelled. While the mainland kingdoms largely kept them out until the nineteenth century, the societies of the island world had to endure their increasingly intrusive presence. For some, especially small-sized and vulnerable societies, the Europeans offered possibilities as powerful protectors. Some scholars speak of the "stranger kings" syndrome, where outsiders are installed inside and accorded governing powers. While outsiders may have frightening aspects, they can also be an asset due to their very foreignness. They may stand detached from local competitors for authority and therefore be in a position to bonding adjudicate in fragmented and strife-torn societies. This might be a reason why certain Southeast Asian societies accepted colonial suzerainty from an early stage despite its oppressive aspects (Henley 2002).

For such societies, the bonding process in a historical past was petrified in stories that were also charters which inscribed the present social order. There are numerous narratives of this kind in the small-scale societies of eastern Indonesia and Timor-Leste, often including marriages that installed the foreigners in the local order. In Larantuka and Sikka in East Flores, a complex of stories tells of early culture heroes who make contact with arriving Portuguese ships and are educated among the foreigners, subsequently introducing Catholicism. While this was evidently a long and uneven process, it is here distilled into particular events with little regard for chronology. Thus, Dom Augustinho, who introduced the Christian creed in Sikka, is represented as the son of the king of Melaka, disregarding the Muslim identity of the real sultans of Melaka before 1511 (Lewis 2010). In Larantuka, the person who first disseminates Catholicism is actually the son of the *tuan tanah* (lord of the land), making the spread of the true faith an indigenous initiative as much as a European one (Heynen 1876, 2–13).

On nearby Timor, the indigenous dynamics in bonding the foreigners are likewise stressed. A story recorded in the nineteenth century tells of a fisherman from Solor, an island east of Flores which kept a Dutch garrison after 1613. When he once sailed his craft to the open sea, a big fish swallowed the hook and dragged the boat all the way to Kupang in westernmost Timor. The curious locals gathered at the seashore and asked where he came from. The fisher replied that he came from Solor, where the Dutch were an ordering force and conditions were advantageous. The local Timorese then decided to invite the white foreigners to establish a post in Kupang, which they willingly did (Heijmering 1847, 46–9). Historically, the coming of the Dutch to Kupang is dated 1653, though the circumstances bear no resemblance to the oral tradition. In this case the colonial establishment is fated through the intervention of the large fish.

Stories like this may raise eyebrows since they so obviously depict the colonials in a positive light as an ordering force. One factor is the acculturation that was going on. On Flores, Solor, and Timor, intermarriage was common, resulting in Christian Eurasian populations in the colonial ports (Hägerdal 2012, 43–7, 133–8, 255–99). Naturally, these stories do not represent a unanimous endorsement of the colonial practices among local societies. On the contrary, there are also numerous narratives from eastern Indonesia about abusive practices on the part of Dutch and Portuguese people who have bonded indigenous kings and elites. In these stories, mendaciousness and base stratagems play a large role. A story recorded in Ambon (Moluccas) in 1678 tells of Sultan Babullah of Ternate, a truly heroic figure who expands the territory of the spice sultanate in various directions. Historically, this individual is well attested to have reigned in 1570–1583 (Andaya and Andaya 2015, 166–7). As the sultan returns from a successful expedition from Sulawesi, a Portuguese ship lies at anchor at Ternate Island and invites him for deliberations. Babullah steps on board and is immediately arrested by the Europeans, who sail to Ambon with their prisoner, before proceeding towards India. A number of Ambonese people manage to come aboard to greet the sultan, having tied daggers to their bodies without the Portuguese noticing. They offer to run amuck against the Portuguese and liberate the ruler. However, Babullah, knowing his fate, refuses their help and later dies during the sea trip to Portuguese Goa (Rumphius 2001, 150). The story is fictional since the historical Babullah died from illness at home, but the framing indicates local feelings towards the rapacious and untrustworthy Portuguese.

Accounts of European-indigenous altercations may also combine untrustworthiness on the part of former with supernatural skills on the part of the latter. The historical Timorese ruler Nai Baob Sonba'i fled from the might of the oppressive Topasses (Portuguese Eurasians) in 1748 and found refuge in Kupang in westernmost Timor where the Dutch maintained a post. He was accepted as a subordinated ally of the Dutch East India Company. However, the ruler was arrested and exiled by the Dutch in 1752 since he was suspected of colluding with the Portuguese, rivals of the East India Company in this part of Southeast Asia (Hägerdal 2009, 57–8). This incident is remembered in local tradition as a contest in magic. The Dutch commander teasingly invites the ruler to prove his innate power by performing a succession of impossible tasks. In fact, the ruler, who thereby puts the Dutchman to shame, wins the various sections of the contest. Nai Baob Sonba'i first turns into a snake, which enables him to crawl through the hole of a blowpipe, then puts a burning wax candle into his mouth without being hurt, and finally outweighs a load of sandalwood—the Timorese export product par excellence. The European commander, fearful of the magic powers of his adversary, eventually resorts to base treachery and exiles the ruler to the colonial capital Batavia (Fobia 1984, 78–81). In this way, the historiographical tables are turned, and the element of treason is transferred from the indigenous lord to the white man.

Explaining European Ascendency

The European political ascendancy in Southeast Asia was a slow process, and it seems likely that most people in the region up to the nineteenth century had never seen a European, although they may well have felt the consequences of Western economic encroachment in a variety of ways. However, as the colonial grip on Southeast Asian territories tightened, there arose a need to explain the process. The foreigners had other customs, other values, and other religious beliefs but still prevailed and forced local elites to yield.

One way to explain all this was that the Westerners after all had gained a degree of legitimacy, although it was perhaps not of the type expected by a Western reader. Javanese historical tradition informs us that an ancient kingdom on the island, Pajajaran, possessed great spiritual power. Although it fell to a Muslim conqueror in 1579, a princess with a flaming womb was issued by the ruling dynasty. Though attractive, her flaming genitals made intercourse impossible for any man, and she was finally banished to an island outside of modern Jakarta. Later on, a certain Dutch captain bought her for three magic cannons. The Dutchman was actually able to sleep with her, and the pair gave rise to the Dutch colonial rulers. For that reason, the Europeans were legitimately established in West Java (Batavia, Jakarta) and allied with the kings of Java (Reid 1994, 292). A variant of this is found in the fantastic romance of Baron Sakender, where the son of the Pajajaran princess and the Dutchman is no other than Jangkung, i.e., the historical empire builder Jan Pieterszoon Coen (1587–1629). Through his bloody exploits in building up Dutch power in the islands, he avenges his insulted mother and the Muslim conquest of Pajajaran (Reid 1994, 293–4). Thus, particular events, which are in themselves ultimately historical, are rearranged in a new chain of causality to provide a rationale of colonial rule.

Another given reason for the Western ascendancy was the foreigners' ability to overcome the magic of their opponent. This is seen in stories about the Dutch conquest of Bali in 1906 and 1908. Powerful heirloom weapons play an important role in Southeast Asian cultures, and military successes are often attributed to these objects. As highlighted by Margaret Wiener, there is/was a Balinese belief that the Dutch managed to neutralize the heirlooms, which had been so efficient in halting the invaders on earlier occasions. This stratagem explained the sudden and complete takeover of the colonial forces, after notoriously mowing down the suicidal attacks (*puputan*) of the old Balinese elite (Wiener 1995). But there is also a sense that the colonial conquest is fated and accompanied by natural phenomena that forebode the disaster. The Balinese chronicle *Babad Dalem* provides a dramatic account of the demise of Klungkung, the senior kingdom of Bali, in 1908. It might be noted that the omens are noted down in some detail while the battle itself is only mentioned in extreme brevity and without much animosity towards the Dutch:

> A long time after His Majesty [I Dewa Agung Gede Jambe] became king, during about two years, there were signs of disorder. There were many signs that foreboded the destruction of the land of Smarawijaya [Klungkung]. Suddenly, the waringin tree that grew in the yard outside the Smarawijaya Palace had white leaves on a branch. And there was an eclipse that was not at the right time. The body of a human without head was seen touching the rays of the sun. The Unda River was also flooded and washed away countless trees. Crying was heard every night in the four junctions of Klungkung town, without it being known where it came from. Such is the story. Then the land of Smarawijaya was attacked by the Dutch troops. Eventually the Dutch troops managed to subjugate the land of Smarawijaya. I Dewa Agung Gede Jambe fell on the field of battle in the *puputan*. (Warna 1986a, 129)

More common, perhaps, is the perceived ability of the Europeans to resort to trickery. There is a comprehensive category of trickster tales in various Southeast Asian cultures, and colonial power expansion is conveniently fitted into such stories. The well-known tale of a local ruler offering a newcomer as much land as he can cover with an ox hide, upon which the newcomer cuts the hide in narrow strips and is able to cover a substantial territory, is found in several places. In a Western context, the story is applied to the founding of ancient Carthage and Viking-age York. The theme is interestingly also located in various parts of Southeast Asia, where it is used to explain European settlements in Melaka, Batavia, and Syriam (Burma) (Reid 1994, 290–1). Again, we see how widespread literary tropes were used to give an edge to the dramatic coming of cultural strangers, whose unpredictability and cunningness are emphasized.

Conclusions

While Southeast Asia is a vast region with multiple historiographic and literary traditions, this brief essay has highlighted some characteristics of the "fantastic" Occidentalism that evolved there. The outlay of all these imaginations depended on a set of factors: the aims of the foreigners (war, conquest, trade, missionizing), the intensity of the relations (temporary visits, frequent intercourse, settlement), and the perceived cultural differences. Some general trends may be discerned across the region.

Southeast Asians, especially in the archipelagic world, were used to multi-cultural milieus. Ethnic and cultural differences between "us" and "them" are sometimes noted in chronicles and literary texts, but more often not, as they seem to be of little concern. "Race" does not play a role in this essay; construction of inside and outside groups is a general human phenomenon but did not translate into the ontological bifurcation tied to power and knowledge production that is seen in modern Western history and occasionally in Asian civilizations (Araújo and Maeso 2015, 22; Dikötter 1992). For most Southeast Asians, there was no way to know if the Europeans had been present for 200 or 400 or 1000 years, and they could even be included in lists of peoples present since ancient times. Moreover, it is striking that historical traditions often tell very little of Europeans even in countries where we know that they were present and had some economic impact. Until the late colonial era, they were simply not important for the day-to-day life of most people. For members of the learned elites, China, India, and Arabia might have more important as points of reference. One may argue that Southeast Asian Occidentalism is different from European Orientalism through the inferiority of power capabilities of the Southeast Asians that would have impeded a Saidian positional superiority vis-à-vis the foreigners, but this is only partly applicable. For much of the period from the sixteenth to the nineteenth century, there was no way that locals in, say, Burma or Sumatra could tell that Europeans would gain political and economic hegemony in the future.

The white foreigners do occur, however, in disruptive and dramatic situations, often relating to war and conquest. They were often depicted as frightening, dangerous, untrustworthy, and cunning. Their behavior could be lewd and despicable, even when religious and cultural differences were omitted in the stories. There is a difference here from other foreign groups, such as Arabs, Indians, and Chinese. At the same time, their frightening aspects could be handled to the advantage of locals when they were invited to protect and adjudicate vulnerable societies, thus performing the role of stranger kings. Whatever the case, the Westerners were rarely players on the same conditions as Southeast Asian ethnic groups. Although they were present since the sixteenth century and although they readily intermarried with local women, they remain essential outsiders in tradition and lore.

References

Andaya, Barbara Watson, and Leonard Y. Andaya. 2015. *A History of Early Modern Southeast Asia, 1400–1830*. Cambridge: Cambridge University Press.

Araújo, Marta, and Silvia Rodríguez Maeso. 2015. *The Contours of Eurocentrism: Race, History, and Political Texts*. Lanham: Lexington Books.

Brown, C.C. (tr.). 1970. *Sejarah Melayu, 'Malay Annals'*. Kuala Lumpur & Singapore: Oxford University Press.

Buruma, Ian, and Avishai Margalit. 2004. *Occidentalism: The West in the Eyes of its Enemies*. New York: Penguin Press.

Carrier, James G. 1995. Introduction. In *Occidentalism; Images of the West*, ed. James G. Carrier, 1–30. Oxford: Oxford University Press.

Dikötter, Frank. 1992. *The Discourse of Race in Modern China*. London: Hurst.

Fobia, F.H. 1984. *Sonba'i dalam kisah dan perjuangan*. Soe. [Unpublished manuscript].

Hägerdal, Hans. 2009. The Exile of the Liurai; A Historiographical Case Study from Timor. In *Responding to the West; Essays on Colonial Domination and Asian Agency*, ed. Hans Hägerdal, 45–68. Amsterdam: Amsterdam University Press.

———. 2012. *Lords of the Land, Lords of the Sea: Conflict and Adaptation in Early Colonial Timor, 1600–1800*. Leiden: KITLV Press.

———. 2017. Timor and Colonial Conquest: Voices and Claims about the End of the Sonba'i Realm in 1906. *Itinerario* 41 (3): 581–605.

Hamidi, Tahrir Khalil. 2013. Edward Said and Recent Orientalist Critiques. *Arabic Studies Quarterly* 35 (2): 130–148.

Harvey, G.E. 1925. *History of Burma, from the Earliest Times to March 1824, the Beginning of the English Conquest.* London: Longmans, Green and Co.
Heijmering, G. 1847. Bijdragen tot de geschiedenis van Timor. *Tijdschrift van Nederlandsch-Indië* 9 (3): 1–62, 121–232.
Henley, David. 2002. *Jealousy and Justice: The Indigenous Roots of Colonial Rule in Northern Sulawesi.* Amsterdam: Free University Press.
Heynen, F.C. 1876. *Het Christendom op het eiland Flores in Nederlandsch Indië.* 's-Hertogenbosch: Van Gulick.
Lewis, E. Douglas. 2010. *The Stranger Kings of Sikka.* Leiden: KITLV Press.
Mak Phoeun. (ed. and tr.). 1981. *Chroniques royales du Cambodge (de 1594 à 1677)* [Collection de Textes et Documents sur l'Indochine, xiii]. Paris: École Française ďExtrême-Orient.
Manusama, Zacharias Jozef. 1977. *Hikayat Tanah Hitu, Historie en sociale structuur van de Ambonse eilanden in het algemeen en van Uli Hitu in het bijzonder tot het midden der zeventiende eeuw.* PhD thesis, Rijksuniversiteit te Leiden.
Ras, J.J. 1968. *Hikajat Bandjar: A Study in Malay Historiography.* The Hague: M. Nijhoff.
Reid, Anthony. 1994. Early Southeast Asian Categorizations of Europeans. In *Implicit Understandings; Observing, Reflecting, and Reporting on the Encounters between Europeans and Other Peoples in the Early Modern Era*, ed. Stuart B. Schwartz, 268–294. Cambridge: Cambridge University Press.
Rumphius, Georgius Everhardus. 2001. *De generale lant-beschrijvinge van het Ambonse gouvernement ofwel De Ambonsche lant-beschrijvinge.* Amsterdam: W. Buizje.
Said, Edward. 1978. *Orientalism.* New York: Pantheon Books.
Spillett, Peter. 1999. *The Pre-Colonial History of the Island of Timor Together with Some Notes on the Makassan Influence in the Island.* Darwin: Museum and Art Gallery of the North Territory. [Unpublished manuscript].
Tarling, Nicholas, ed. 1999. *The Cambridge History of Southeast Asia, Volume One, Part Two: From c. 1500 to c. 1800.* Cambridge: Cambridge University Press.
Teeuw, Andries, et al. 2004. *A Merry Senhor in the Malay World; Four Texts of the Syair Sinyor Kosta.* Vol. II. Leiden: KITLV Press.
Traube, Elizabeth. 1986. *Cosmology and Social Life: Ritual Exchange Among the Mambai of East Timor.* Chicago: University of Chicago Press.
Warna, I Wayan. 1986a. *Babad Dalem: Teks dan Terjemahan.* Dinas Pendidikan dan Kebudayaan, Provinsi Daerah Tingkat I Bali.
———. 1986b. *Usana Bali; Usana Jawa: Teks dan Terjemahan.* Dinas Pendidikan dan Kebudayaan, Provinsi Daerah Tingkat I Bali.
Wiener, Margaret J. 1995. *Visible and Invisible Realms: Power, Magic, and Colonial Conquest in Bali.* Chicago: University Chicago Press.

CHAPTER 6

How [Not] to Run a Colony in the Distant Past and the Future

Karen Ordahl Kupperman

Any venture to create some kind of society in an unknown land, whether four or five centuries ago or in the future predicted by science fiction, must have a model in mind for how the colony will operate. But imaginary models and reality are usually quite different. Managing relations with the inhabitants of the new place will always be a major concern; however the travelers regard those residents and their rights. Even more challenging is managing the people who are part of the proposed colonial effort, especially once they arrive at their new home. Given that this environment is a novel and probably very challenging land in which inherited knowledge and therefore rank become irrelevant, management is crucial. And very difficult. Those who seize command attempt to create a structure in which everyone's behavior is strictly controlled and hierarchy is maintained.

As we look at these ventures, one key question is "why are they leaving their homeland?" In science fiction accounts, such as *Aniara*, the epic poem by Harry Martinson, written in 1956 in 103 Cantos, tells of a spaceship carrying "eight thousand souls" from the region of Doris land on earth to Mars "when Earth, become unclean with toxic radiation," is no

K. O. Kupperman (✉)
New York University, New York, NY, USA
e-mail: karen.kupperman@nyu.edu

J. L. Hennessey (ed.), *History and Speculative Fiction*,
https://doi.org/10.1007/978-3-031-42235-5_6

longer habitable (Martinson 1999 [1956], Canto 2). In founding Jamestown, English leaders wanted to establish a foothold in America. But they were also concerned about the perception that England was overpopulated. As Robert Gray addressed the Virginia Company in 1609:

> There is nothing more daungerous for the estate of common-wealths, then when the people do increase to a greater multitude and number then may justly paralell with the largenesse of the place and countrey: for hereupon comes oppression, and diverse kinde of wrongs, mutinies, sedition, commotion, & rebellion, scarcitie, dearth, povertie, and sundrie sorts of calamities … For even as bloud through it be the best humor in the body, yet if it abound in greater quantitie then the state of the body will beare, both indanger the bodie, & oftentimes destroyes it. (sig. B3v)

Offloading superfluous population was one justification for starting English colonies across the Atlantic, and the first people sent over were young men who needed employment.

Moving ahead, the key tools for colonial projects are language and memory. The first tests occur on the voyage, which is inevitably full of pitfalls. Crossing the Atlantic or venturing into space can be subject to all kinds of unknown accidents: storms can push vessels way off course; unknown dangers such as objects in space, or underwater reefs or projections can wreck the best-laid plans. Shakespeare's (1623) *Macbeth* was first performed in 1606, just at the time Virginia Company investors were deep in preparations for founding their colony, Jamestown, in what the English called Virginia. When the witches in the play decided to torment the master of the *Tiger* and his crew, keeping them tempest-tossed for nine times nine weeks, they were referring to an event of which the audience would have been keenly aware: the recent arrival of a ship named the *Tiger* that had indeed been tossed by storms and sent wandering for 567 days, or nine times nine weeks (I:3; Loomis 1956, 457). John Donne's poem "The Storme" described a sea storm he had lived through; the tempest was so ferocious that "Compar'd to these stormes, death is but a Qualme, Hell somewhat lightsome, and the' Bermuda calme." (Donne 1633, 56–9)

The storm and shipwreck that opened Shakespeare's *Tempest* were based on reports from the large fleet of five ships the Virginia Company sent to replenish the colony in 1609 whose flagship, aptly named the *Sea Venture,* was caught in a huge hurricane whose "winds and seas were as mad as fury and rage could make them" and wrecked on Bermuda, which was known as the Isle of Devils because it was surrounded by dangerous underground reefs (Strachey 2013 [1609], 7). As John Donne noted, many ships had wrecked on those reefs and the islands were uninhabited.

The Goldonder Aniara of Martinson's poem was on a voyage "doomed to be a space-flight like to none." The ship swerved to avoid the Honda asteroid "(herewith proclaimed discovered)," which put them in "dead space" outside inner space and the orbits of the planets. "That was how the solar system closed its vaulted gateway of the purest crystal and severed spaceship Aniara's company from all the bonds and pledges of the sun." The emigrants continued to send out signals, but "our call-sign faded till it failed: Aniara" (Cantos 2, 3, 4).

Other ships in the 1609 fleet finally made it to Jamestown, but seawater had flooded the holds and ruined all the supplies they carried. This was especially disastrous because that area of Chesapeake Bay was going through the worst drought in more than 770 years. The drought had started in 1606 and would continue through 1612 (Blanton 2000, 74–81). Thus, the winter of 1609, with more people and no supplies from England, was the infamous Starving Time, with dramatic population loss.

Those who had been wrecked on Bermuda built two new ships and finally arrived in Jamestown almost a year later; they were appalled to see the chaos inside the fort. Everyone was loaded onto the Bermuda-built ships to return to England, but they encountered a new fleet and governor from England coming up the James River as they were traveling down. So they all went back upriver. New government and new supplies were insufficient for stopping the horrible apathy; getting the drones working was still an aspiration, not a reality. Sir Thomas Dale arrived in Jamestown in May 1611, and he was disgusted to find the colonists at "their daily and usuall workes, bowling in the streets" (Hamor 1615, 27–8). When the new governor, Sir Thomas Gates, arrived soon after, he and Dale set about correcting the problem. In response to the horrifying situation, the new leadership imposed the most extreme code of laws in 1610–1611 (Strachey 1953 [1612]).

On working days, everyone was required to exit their homes together "at the first tolling of the bell" to go to church and again when the bell tolled at the end of the day (Law 22). Every soldier and tradesman, which meant almost all men in the colony, was "upon the beating of the Drum to go out to his work" and was not to leave work "before the Drum beat again" (Law 28). When men were sent by boat to explore into the interior, desertion or taking the boat to a different port meant death (Law 32).

Everyone had to be examined by the minister about their faith and to receive instruction by him if their faith was not firm or correct (Law 33).

Cleanliness was a huge concern because of the common belief that mal air produced by putrefaction caused disease. Launderers who washed "foul linen" inside the fort or threw dirty water or suds into the street, and anyone who did "the necessities of nature" in the streets was to be whipped for these "unmanly, slothful, and loathsome immodesties" and turned over to the martial courts for further punishment (Law 22). Everyone was required to keep their residence "sweet and clean" and to make sure their bedstead was three feet off the ground (Law 25).

Death was the punishment for many infractions, including blasphemy (Law 3), murder, sodomy or rape, sacrilege or stealing, bearing false witness (Laws 8, 9, 10, 11), trading with the Native people without permission or stealing from them (Laws 15–16), embezzlement of supplies by those entrusted with them (Law 17), killing any domesticated animal without permission (Law 21), running away to "any savage Werowance" (Law 29), disobedience of authority or treason (Law 30), and "wilfully" plucking any root, herb, flower, or any grapes or ears of corn in a garden (Law 31).

A cook or baker who provided less cooked food than the supplies they had been given would lose both ears; those doing it a second time were condemned to galley slavery for a period of years. Several offenses were similarly punished, with an increasing tier of punishments ending in galley slavery, including failure to attend church, cursing, slander, and disobedience. As leading colonist Ralph Hamor wrote, these laws were harsh, but necessary (Hamor 1615, 27–8).

Colonial leaders and backers in England drew on several models as they constructed their concept of a well-functioning society. One, popular since ancient times, was from nature: the beehive. The Roman author and statesman Pliny the Elder's work was translated into English in 1601 as *The Historie of the World: Commonly Called the Natural Historie of C. Plinius Secundus*. Many colonial backers would have read the *Natural History* in Latin, but its publication in English now meant that its content was more widely available and presumably in demand. Pliny considered bees "the chief insect." Bees "endure toil, they construct work, they have a government and individual enterprises and collective leaders, and, a thing that must occasion most surprise, they have a system of manners that outstrips that of all the other animals."

As under the Jamestown code, bees all slept together. "They sleep until dawn, until one bee wakes them up with a double or triple buzz as a sort of bugle-call; they all fly forth in a body, if the day is going to be fine." People watched to see if they left the hive at dawn, because if they stayed in, winds and rain were coming. At the end of daylight, buzzing within the hive declined "till one bee flies round as though giving the order to take repose with the same loud buzz with which she woke them, and this in the manner of a military camp: thereupon they all suddenly become quiet."

During the day, bees keep "wonderful watch on the work in hand." Work is equally divided; in the hive "some build, others polish, others bring up material, others prepare food from what is brought to them; for they do not feed separately so that there shall be no inequality of work or food or time." Also, "They are wonderfully clean," and extremely thrifty (Pliny 1938–1963, XI, 439–41, 445–9, 473).

Captain John Smith became the President in Jamestown's beginning after all the elite men became incapable of serving. As he wrote, "At this time were most of our chiefest men either sicke or discontented, the rest being in such dispaire, as they would rather starve and rot with idlenes, then be perswaded to do anything for their owne reliefe without constraint" (Smith, 1986a [1608], I:35). As president he instituted his famous program that put everyone, even the most elite, to work: no work, no food.

Smith drew on the bees' example: "If the little Ant, and the sillie [simple] Bee seek by their diligence the good of their Commonwealth; much more ought Man. If they punish the drones and sting them steales their Labour; then blame not Man. Little hony hath that hive, where there are more Drones then Bees: and miserable is that Land, where more are idle then well imployed" (Smith, 1986b [1616], I:311). Pliny wrote that drones were imperfect bees, and many in England scornfully labelled the Jamestown colonists the "the very scumme of the land" (Winthrop 1629, 142–3).

Smith was badly injured when his powder bag exploded in his lap while he traveled up the James River in a boat, and he returned to England on one of the ships that survived the 1609 hurricane (Smith, 1986c [1624], III:223–4). Thus, he was not present for the starving time and the institution of the *Lawes Divine, Morall, and Martiall, etc.* Those laws were extreme, but enforcement was episodic, and many people found their own paths. Jamestown saw colonists leaving to live with the Chesapeake region's Native People. Many traded across ethnic lines; in fact, such trade was essential to keeping the colony alive. The death penalty, although specified for a large number of infractions, did not make sense in such a small and depleted colony. And as far as maintaining strict religious conformity goes, we know that Roman Catholics, some spying for Spain, lived in the colony (Goldman 2011).

Backers interested in the region to the north also cited bees' methods as a model for their enterprises. Richard Eburne, promoting interest in Newfoundland, argued that planners must send families and people of all ages: "Look but into the beehives when they swarm and you shall find … that the swarm is as old as the stock—that is, that there are in it as well old bees as young." In a marginal note, he cited Charles Butler's (1600) *The Feminine Monarchy: or, A Treatise Concerning Bees* (Eburne 1962 [1624], 142). Answering the claim that the families in New England might be vulnerable to a Spanish invasion, William Wood replied: "… when the Bees have honie in their Hives, they will have stings in their tailes" (Wood 1634, 54). But a few years in, the New England colonies were having their own drone problems as noted by Rev. Nathaniel Ward of Ipswich, Massachusetts. Ward wrote to John Winthrop, Jr. complaining of the "multitudes of idle and profane young men, servants and others," burdening the colony, so that parishioners were complaining "with greif we have made an ill change, even from the snare to the pitt" (1635, II: 215–7).

Also, the demand for religious control over life in New England led to a plethora of exoduses. By the middle 1630s, five years after the puritan colony in Boston was founded, colonists began to object to the fact that only those approved by the congregations had full religious and civil rights. Followers of Anne Hutchinson, who openly challenged the ministers' teachings as puritans had done in England, moved to found a new colony in New Hampshire. Roger Williams and Thomas Hooker, both eminent ministers and scholars, moved with their congregations to the future Rhode Island and Connecticut, where they enacted complete separation of church and state. Williams, who had objected to the display of royal symbols in Boston, was under sentence of death if he attempted to return to Massachusetts Bay.

* * *

Modern science fiction returns us to the problem of managing migrants whose expectations are frustrated and who can veer out of control, especially as they have no home to go back to. Aniara's passengers left a planet under Control Plan Three after the end of the Thirty-second War. On Earth, punch cards had replaced all other kinds of relations. As a result, "everyone was playing at least four roles in political games of specters' peek-a-boo." Resisters had been sent to work camps on Venus and Mars, where life was extremely tough (Canto 15). The Space-Hand's Tale, Canto 40, tells how he and his colleagues took "something like three million frightened people to their current star" over five years. Some were sent to ice-cold Mars, and others to the marshes of Venus. One woman, Nobby, who was herself crippled by radiation, maintained the good of humanity by carrying supplies to the exiles, and cleaning and sewing for them. Nobby had her own language: song. The Space-Hand remembered her as "Nobia the Samaritan" (Canto 40).

After the ship veered into endless space, those on board tried to keep life in line with customs on earth so, although all they saw outside the ship was darkness, they decreed periods of day and night based on what they believed was the reality at home as they drifted into eternity. On Midsummer night, they danced all night waiting for the sun to come up, "Then smack came clarity, the horror that it never did come up." They realized that everything was just a dream (Canto 7). Aniara's passengers came to realize that the space in which they traveled was not space as they envisioned it on earth. Actually, they were "lost in spiritual seas" (Canto 13).

"The sterile void of space is terrifying," (Canto 10) but the ship carried an AI installation called Mima, and Mima offered passengers a sense of life and security by pulling in "traces, pictures, landscapes, scrapes of language . . . Our faithful Mima does all she can and searches, searches, searches" (Canto 6). Mima offered a world that blotted out both the reality of the wandering spaceship and the reality of the dying earth they left behind, and people began to worship Mima "as a holy being" (Canto 7). The inventor who created Mima realized that he did not comprehend part of her—the part she built herself (Canto 9).

Mima's output became erratic. At one point, she began showing images of the dying earth (Canto 19). Soon Mima, "blinded by a bluish bolt" showed the destruction of Dorisburg, which shattered the narrator who was also the person who tended Mima, the Mimarobe, Mima's "faithful priest in blue" (Canto 26). Mima begged for relief, and she recovered in five days, but on the seventh day, she emitted a strange drone and she talked in "higher ultramodern tensor-theory" and told the Mimarobe that she had been hearing cries from Dorisvale for some time. Now her "cell-works" were darkened by human cruelty, and she was decaying (Canto 28). As people rushed in to receive her comfort, a "bolt-blue light" flashed from her screens with a loud thunderclap. Many were crushed to death fleeing the hall as Mima died. Her last message was from The Detonee who acknowledged how difficult it was as he detonated her. Everyone blamed the Mimarobe for this horrible failure (Cantos 29, 30).

One passenger, Daisy, continued to speak as if she was still on earth and refused to give in to sadness; in her language, she retained the memory of their earlier life and "slings at Death's void the slang of Dorisburg." The narrator found relief in Daisy's "womb of hair" (Cantos 12,13). After Mima showed the destruction of Dorisburg, Daisy was the "only comfort left" because she still spoke the city's language and the narrator was the last man who understood her as she "babbles in her lovely dialect." As time goes on and on, Daisy is in permanent sleep and dancing is forgotten (Cantos 27, 72). A blind woman from the planet Rind who had chosen to come aboard Aniara created poems and sang songs from Rind. She argued it was "right to witness a delight in living," but listeners said her poems were merely words and wind (Cantos 48, 49).

After Mima died, Chefone, "the fierce lord of our craft," took control: "he struck you as a man out to compel his people to decline and asininity" (Canto 30). The narrator and others, including the technicians who were trained in the "Fourth tensor reportory," were imprisoned deep in the ship's bowels. From their cells, they tried to explain what had happened in ever simpler language and finally in drawings. After studying Mima's language cycles for three years, the narrator was taken up to try to revive her. But without her guidance, he reeled, realizing that her interior was cold. New religions featuring priests, temple bells, and crucifixes, "vagina-cult and shouting yurgher-girls and tickler-sectaries forever laughing" took over Mima's space (Cantos 31, 32, 33, 35).

Colonial ventures that actually landed—whether at their original destination or not—faced a panoply of new problems. In Charlie Jane Anders's *The City in the Middle of the Night* (2019), the migrants had arrived, many generations ago, on the planet January, which is tidally locked, so that one side has permanent sunlight, a scorching environment intolerable to humans.[1] The other side is in complete darkness and is inhabited by indigenous beings whom the humans saw as simple-minded and hostile. The human arrivals settled their city of Xiosphant in the twilight between the two zones, where they were protected from extremes by two mountains and the light was always the same. Huge farmwheels reaching into the sky grew their food.

The ship on which they arrived had separate compartments for people from different cultures on earth, thus maintaining old hierarchies, but on their arrival on this strange planet, those at the top "had no clue how to cope." The monotony of life in the twilight threatened this hereditary leadership, so elites dealt with it and the danger of uprisings as in Jamestown by attempting to regulate every aspect of life. Even the Xiosphanti language embeds the speaker's and listener's status. In a system called Circadianism, shutters on all windows go up at the same time and everyone goes to sleep; shutters come down when it is time to wake up, and immediately the streets are filled with people rushing to their assigned jobs. As the elite student Bianca tells The Progressive Students Union, "If you control our sleep, then you own our dreams . . . And from there, it's easy to master our whole lives." Bianca has access to the college library and is a great reader. She has found a book in the rubble at the back of the shelves containing the history of the first arrivals on January and she tells her fellow students: "That's what history is, really . . . the process for turning idiots into visionaries" (Anders 2019, 19, 15).

Security patrols make sure that no one is outside at the wrong time, and any violation of assigned roles meets extreme punishment. Near the novel's beginning, one of the principal characters, a low social-status student named Sophie, is arrested and dragged up a mountain, where she is cast to her certain death on the dark side for having been discovered with three stolen food credits. Actually it was her best friend, Bianca, who had stolen the credits almost as a joke. Bianca belonged to one of the most favored dynasties on January, and Sophie picked them up to shield her when the authorities arrived.

As in Jamestown, regulations are one thing, but absolute control is quite another. In Xiosphant, most people were caught in the endless cycle of work and sleep, but throughout *The City in the Middle of the Night,* we encounter people who have exited that system, or who, like Bianca, are able to ignore it to some extent. The city of Argelo beyond the tempestuous, icy Sea of Murder, is the exact opposite of Xiosphant: a place where anything goes, The City that Never Sleeps (170). It is not necessarily a great place to live, but everyone is free and they have their own language. Groups like the Resourceful Couriers travel across the twilight, carrying illegal goods from place to place and supporting life inside and outside the system.

Relationships with indigenous populations are always a huge part of these systems, but one that no one in charge wants to talk about. Sophie was thrown to her death on the dark side, but she did not die. All the humans who lived on January, whether in the monotony of Xiosphant or in the wild freedom of Argelo, believed that the planet's original inhabitants, called crocodiles by the humans, were stupid, vicious animals, who might be threatening and who had no value except when they were hunted as food. Sophie quickly absorbs a wholly different reality when she lands at the foot of the mountain on the dark side. She is saved by one of the "crocodiles," whom she calls Rose; her new friend warms her and communicates through tendrils on her chest. As Rose opens her chest, Sophie puts her face among the warm tendrils as the Mimabore had put his face in Daisy's welcoming mop of hair (31). Sophie tries to use the name they call themselves, which she renders as Gelet, and on various trips bringing needed supplies to Rose she begins to understand the amazing complexity of the technology on which their underground City in the Middle of the Night runs. She is given an electronic bracelet through which the Gelets communicate with her wherever she is. She learns that at the beginning, the Gelets had considered wiping out the humans before they could get their city of Xiosphant fully established. But then they realized humans were slowly losing contact with the Mother Ship and could be useful to them, especially in providing essential metals from mines and meteors (312–3).

Because she cannot live openly in Xiosphant after she returns, Sophie gets to know people who live a different life even within the city. As in Jamestown, many individuals found ways to escape the rigid rules. She remembers a place her mother took her to and she reconnects with Hernan, who runs a refuge there for people like her. She meets other outcasts, Mouth and Alyssa, and travels with them and their friends to Argelo. Mouth is the last of her nomadic people, the Citizens, who spoke a language called Noölang, and she is desperate to retrieve a document, The Invention, that contains her people's histories and ceremonies so she can understand what she has lost. The document is in the palace in Xiosphant, and she is enlisting allies to get access to it even when she is in Argelo. Maybe Bianca will help.

Memory, and the language in which it is preserved, is crucial to all these efforts. In Argelo, Mouth meets an academic who has spent a lifetime studying the Citizens. He is eager to talk to her, but is it possible that he knows more about them than she does? To whom does memory belong? He introduces her to a former Citizen who had decided to leave the group and thus escaped the slaughter in which only the child Mouth was left alive.

Names are crucial in all these stories. Mouth did not actually have a name because she had not yet arrived at the level to be given one when her people were wiped out. Sophie named her rescuer Rose, and her elite friend, who became more and more distant as their paths diverged, was named Bianca. The technicians imprisoned after Mima died were stripped of their names (Martinson 1999 [1956], Canto 34). In essence, they were erased until their services were needed again.

Pocahontas had many names; Pocahontas was actually a nickname describing her playfulness. Her sacred name was Matoaka; her regular name was Amonute. When she was baptized with the important biblical name Rebecca by the puritan minister Alexander Whitaker, who also presided over her marriage to John Rolfe, her sacred name was revealed. Samuel Purchas, who interviewed returning venturers and collected their documents, wrote that her people had feared the English could harm her if they knew her sacred name (see Wood 2016, 77; Purchas 1617, 943; Whitaker 1615, 59–60).

Colonists everywhere in the Americas assigned names to the land in order to indicate imagined possession and control. Those on the first voyage to the area on the Carolina coast where the ill-fated Roanoke colony would be founded in the 1580s asked the name of the land and thought they had been told Wingandacoia, which they reported back to England. Sir Walter Ralegh, the colony's backer, wrote that the colonists later learned the Natives were commenting on their clothing; actually, we now know that the informants thought the newcomers were pointing to some spruce trees. Wingandacoia appeared on several maps, but the Queen gave Ralegh authority to change the region's name to Virginia in her honor (Quinn 1955, I:116–7, 147, 174, II:853–4).[2]

A satirist, John Healey, played on this mistake. In 1609, he published anonymously what he said was a "translation" of Joseph Hall's more serious *Mundus alter et idem* (1605). The mythical traveler in his *The Discovery of a New World, or A Description of the South Indies. Hetherto unknowne* visited "The new discovered Womandecoia (which some mistaking both name and nation) call Wingandecoia, & make it a part of Virginia)[sic] otherwise called Shee-landt" (Healey 1609, 96 [misnumbered 66], 97).

Healey's jest pointed to a deep concern. English commentators lamented new fashions coming in from Europe and America; men wearing them adopted manners and gestures that marked them as effeminate. Men spending more and more time in fashionable company seemed to yield too easily to female notions. Some said these modish new ways rendered

Englishmen cowardly and incapable of properly defending the nation. Keith Thomas labels this disdain "Xenophobic Masculinity." And moralists were all too eager to blame women for men's decline into effeminacy (Thomas 2018, 219–23).

The *Lawes Divine, Morall, and Martiall* were designed to prevent colonists' degeneration. Partly they were concerned about the men going native and the laws tried to control all contact. But in America as in England, leaders feared the young men would become feminized through their own behavior. As Alexandra Shepard puts it, describing early modern English youths: "In their bids for manhood, young men embraced precisely the kinds of behaviour—violent disruption, excessive drinking, illicit sex—condemned by moralists as unmanly, effeminate, and beastlike" (Shepard 2003, 29, 78–9, 83, 94). In the *Lawes* Strachey discussed regulating the conduct of soldiers in Virginia, saying chastity was absolutely necessary, "when uncleanesse doth defile both body and soule, and makes a man stinke in the nostrils of God & man, and laieth him open to the malice & sword of his enemy, for commonly it makes a man effeminate, cowardly, lasie, and full of diseases" (Strachey 1953 [1612], 81–2). The only way to maintain true English masculine identity was to stay as far away from Others as possible.

As they assimilated Natives' land and sometimes cut them off from waterways, seventeenth-century English newcomers to North America told themselves that their bringing the gift of Christianity more than made up for any losses the original inhabitants suffered. In all cases disdain for original inhabitants mixed with extreme fear meant that settlers lost crucial opportunities to learn about resources and environments. And this was true even when they knew such instruction would be invaluable.

In *The City in the Middle of the Night*, Sophie, as she increasingly spent time with the Gelets and learned from them, learned that they had designed the planet and worked over many generations to perfect it. Without their work, the humans who landed on January could never have survived, but the humans, who manufactured memories that suited them and their needs, could not understand or acknowledge their dependence (Anders 2019, 320).

The Gelets on January retained all memories in their tendrils. And these memories were in the same moment, as if they had just happened. Sophie learns about this way of remembering as she spends more and more time with them, and in the end, she agrees to undergo an operation that gives her tendrils and allows her to understand that this is the way to achieve peace among all beings.

English colonists in North America also created memories, official and informal. The Jamestown laws dictated that "euery Minister shall keepe a faithfull and true Record, or Church Booke, of all Christnings, Marriages, and deaths of such our people, as shall happen within their Fort, or Fortresses, Townes or Towne at any time" (Law 7). Similar official records were kept in New England towns.

We cannot rely completely on written texts because they are not transparent; they are all written to serve a purpose. Writers create a version of events that they hope will create a desired response. This is especially true of reports from Jamestown where writers were afraid that investors in England might just give up on the project and desert them as Ralegh had deserted the Roanoke colonists. Leaders tried to project an immediate future in which everything was just about to improve.

But other sources told a different story. Returning colonists, for one, could not be silenced. As historian Wesley Frank Craven writes, "By 1612 the adventurers [investors] were complaining that only the name of God was more frequently profaned in the streets and market places of London than was the name of Virginia" (Craven 1970, 26–7). Don Diego de Molina, one of the people the Spanish had inserted into Jamestown, was allowed to send home unsealed letters; unsealed meant that Jamestown's leaders could read them first. What those leaders did not know was that Don Diego also smuggled out a truly frank discussion of what was actually going on there for the king of Spain; it was hidden inside the sole of a Venetian gentleman's shoe (Molina 1890 [1613], 646–652; see also Wright 1920). The truth will come out, no matter how hard elites try to keep it hidden.

On January, Mouth, who was so eager to know her people's story and history, learned an awful truth. Through her friendship with Sophie, Mouth had contact with Gelet society and ultimately learned that the Gelets had destroyed the Citizens because they had systematically stripped the mountain of a plant the Gelets had tended over many generations and through which they controlled the skies and made them stable. Once it was gone, the skies became ferocious and a mountain of ice with noxious rain inside it landed on a nursery with thousands of new-born Gelets, killing most and leaving the survivors horribly maimed (Anders 2019, 303–5). As she slowly recovered from the shock, Mouth realized that the Citizens had made up a set of fables about the Gelets. "The Citizens had stayed blameless in their own cosmology, until the very end" (319). But, despite her enlightenment, she still could not bear to attain true memories by undergoing the operation Sophie had chosen.

Jean, one of the infants who survived the storm attack on the nursery, had lost the use of her left tentacle and was unable to do much physical work, so she became a teacher and was good at it. She spent time with Sophie to teach her about their life, and she recounted her memory of the "violent snowfall" in which "the snow grew teeth" as rocks fell on the nursery. She survived because she was under a rock, but her left side was always racked with pain (312–3). Through her stays with the Gelets, Sophie, like Mouth, had also come to understand that humans rely on their power to forget and to make memories that conform to their self-image. She remembered Bianca telling her "progress requires us to curate the past, to remove from history things that aren't 'constructive'" (312).

Those trapped on Aniara had access to good memories as long as Mima continued to function but were anxious when she showed the truth, and bereft once she broke down. Aniara had a Memory Hall, but it was a place where people went to recant their sins and confess their misdeeds wearing ashes on their heads (Canto 59). Chamber Seven on Aniara contained the "files of Thinking" overseen by the "Friend of Thought." The Friend pointed to many ways of thinking that would have prevented the disasters earth had suffered, but it was too late. Occasionally a passenger would come in to look up a line of thought, but interest soon faded (Canto 44).

And some memories were kept hidden or were shared by a few. Mima's "faithful priest in blue" also made reference to the Blue Archive, but not to where it was located or what it might contain (Canto 14). Nor does the reader know what the constant references to the color blue, both as positive and as an agent of destruction, really mean. On January, people with access could see memories of earth in the form of films on the mother ship's computer, but to what end? (312)

The Virginia Company and later colonists worked to create the story of Virginia they wanted. And for many generations historians created a colonial past in which the Native people were savages who did not use the land properly, and the colonists hard-working and generous. Even in the twenty-first century, Indigenous Americans still fight for proper inclusion in that history. And our history is once again being carefully curated to tell the story some want us to hear.

What if management of both people and memories was not the central problem? Some people at the time English colonies were being founded dreamed of a solution to conflict that would be as successful as that achieved by the Gelets. What if music was the way to universal understanding? In Virginia, Native leaders often used music in greeting: George Percy, the Earl of Northumberland's brother and a man used to aristocratic display, described the Rappahanna chief's self-presentation. He came "playing on a Flute made of a Reed, with a Crowne of Deares haire colloured red, in fashion of a Rose, fastened about his knot of haire and a great Plate of Copper on the other side of his head, with two long Feathers in fashion of a paire of Hornes placed in the midst of the Crowne" (Percy 1969 [1606], 135–7).

Seventeenth-century sci-fi described moon voyages and the knowledge voyagers brought back. These writers drew on Matteo Ricci's discussion of the Chinese language. Chinese consists of thousands of characters, each of which has a specific meaning, thus eliminating much of the confusion and danger created by alphabetic languages. Saying thousands of characters is impossible, so spoken Chinese is tonal; early modern Europeans took this to mean it was sung (Ricci 1942–1953 [1583–1610], chap. 5, esp. 26–30; Cornelius 1965; Spence 1984).

Church of England bishop Francis Godwin wrote of the "speedy Messenger" Domingo Gonsales's trip to the moon by way of a swan-powered machine he had invented. Gonsales reported that the lunar people's language consisted mainly of musical phrases. In this system a small number of words sufficed because they were sung in different tunes and thus were able to convey a wide range of meanings, and all communications were clearly understood (Godwin 1638, 93–5, 123).[3] The French writer Cyrano de Bergerac wrote that in lunar society ordinary people communicated by gestures, but the elite conversed in musical tones. For very complicated discussions, lunar gentlemen used musical instruments such as lutes. "[S]o that sometimes Fifteen or Twenty in a Company, will handle a point of Divinity, or discuss the difficulties of a Law-Suit, in the most harmonious Consort, that ever tickled the Ear" (de Bergerac 1687, 43).[4]

Ben Jonson's masque *News From the New World Discovered in the Moon,* performed before King James's court at Whitehall in 1620, discussed a recently returned moon voyage. The Chronicler asks what language is spoken there. The Second Herald replies: mostly by gestures, but they do have "certain motions to music. All the discourse there is harmony." So the seventeenth century had aspirations for universal peace through mutual understanding (Jonson 2014 [1620]).

As we do in modern times. In 1977, NASA sent a Voyager craft into space with the Golden Record that contained many different kinds of earth music. Carl Sagan and the others involved in the project believed that music was the best way to communicate with beings whose life and culture were utterly unknown. As Daniel K. L. Chua and Alexander Rehding write in *Alien Listening*, about "Sagan's vision for Voyager: for him, aliens are strangers that we somehow already know, because we believe the music that bears our signature will resonate with them and trigger an estranged recognition of a garment woven in our time and our place, but now radically untimed and displaced" (Chua and Rehding 2021, 203). The diagram on the cover was designed to help those beings set up a record player with the attached pieces. The spaceship left our solar system in 2012; as on Aniara, Voyager I continues to send faint signals back to earth, but they are harder and harder to receive. The ship's generators will shut down in 2030; after that, like Aniara, it will continue drifting further and further into space. If a being does discover the Golden Record, it is interesting to speculate about the reaction it will elicit (Ferris 2017). In what they call their TARDIS-like book, *Alien Listening*, Chua and Rehding wonder if the final piece, Beethoven's Cavatina, might be "perceived by some alien behemoth as nothing more than a sharp, pertinent fart" (Chua and Rehding 2021, 29, 78–9, 83).

Notes

1. Charlie Jane Anders is my daughter.
2. The name Virginia was applied to North America's east coast, from Florida to Maine.
3. I thank Jennifer Egloff for bringing this example to my attention. On the theological implications of the many discussions of the possibility of life on the moon in the seventeenth century, see Cressy 2006. Cressy argues that 1638 was "England's lunar moment" (967).
4. Cyrano de Bergerac's satire was translated into English twice in the seventeenth century. The first translation, by Thomas St. Serf, was published in 1659 and was unpaginated. This quotation is taken from the second translation, by A. Lovell (1687).

References

Primary Sources

Anders, Charlie Jane. 2019. *The City in the Middle of the Night.* New York: Tor Books.

Butler, Charles. 1600. *The Feminine Monarchy: Or, A Treatise concerning Bees.* London.

de Bergerac, Cyrano. 1687. *The Comical History of the States and Empires of the Worlds of the Moon and Sun.* Translated by A. Lovell. London: Henry Rhodes.

Donne, John. 1633. *Poems By J.D. with Elegies on the Authors Death.* London.

Eburne, Richard. 1962 [1624]. *A Plain Pathway to Plantations*, ed. Louis B. Wright. Ithaca, NY: Cornell University Press.

Godwin, Francis. 1638. *The Man in the Moone.* London.

Gray, Robert. 1609. *A Goodspeed to Virginia.* London.

Hall, Joseph. 1605. *Mundus alter et idem.* London.

Hamor, Ralph. 1615. *A True Discourse of the Present Estate of Virginia.* London.

Healey, John. 1609. *The Discovery of a new world, or A Description of the South Indies. Hetherto unknowne, By an English Mercury.* London.

Jonson, Ben. 2014 [1620]. News From the New World Discovered in the Moon. A masque, as it was presented at court before King James. In *The Cambridge Edition of the Works of Ben Jonson*, ed. Martin Butler. Cambridge: Cambridge University Press.

Martinson, Harry. 1999 [1956]. *Aniara.* Translated by S. Klass and L. Sjoberg. Brownsville, OR: Story Line Press.

Molina, Don Diego de to Don Diego de Velasco. 1890 [1613]. Letter sent May 28 1613. In *The Genesis of the United States*, ed. Alexander Brown, II, 646–652. Cambridge, MA: Riverside Press.

Percy, George. 1969 [1606]. Observations gathered out of a Discourse of the Plantation of the Southerne Colonie in Virginia by the English. In *The Jamestown Voyages Under the First Charter, 1606–1609*, ed. Philip L. Barbour, vol. 1. Cambridge: Hakluyt Society.

Pliny the Elder. 1938–1963. *Natural History.* Translated by H. Rackham, W.H.S. Jones, and D.E. Eichholz, vol. XI. Cambridge, MA: Harvard University Press; London: Heinemann.

Purchas, Samuel. 1617. *Purchas His Pilgrimage.* 3rd ed. London.

Quinn, David Beers, ed. 1955. *The Roanoke Voyages, 1584–1590.* London: Hakluyt Society.

Ricci, Matteo. 1942–1953 [1583–1610]. *China in the Sixteenth Century: The Journals of Matthew Ricci, 1583–1610.* Translated by Louis J. Gallagher. New York: Random House.

Shakespeare, William. 1623. *Macbeth.*

Smith, John. 1986a [1608]. A True Relation of such Occurrences and Accidents of Noate as Hath Hapned in Virginia 1608. In *The Complete Works of Captain John Smith*, ed. Philip L. Barbour, I:35. Chapel Hill: University of North Carolina Press.

———. 1986b [1616]. A Description of New England. In *The Complete Works of Captain John Smith*, ed. Philip L. Barbour, I:311. Chapel Hill: University of North Carolina Press.

———. 1986c [1624]. The Generall Historie of Virginia, New-England and the Summer Isles. In *The Complete Works of Captain John Smith*, ed. Philip L. Barbour, III:223–224. Chapel Hill: University of North Carolina Press.

Strachey, William. 1953 [1612]. *The Historie of Travell into Virginia Britania*, ed. Louis B. Wright and Virginia Freund. London: Hakluyt Society.

———. 2013 [1609]. A True Reportory of the Wreck and Redemption of Sir Thomas Gates, Knight. In *A Voyage to Virginia in 1609*, ed. Louis B. Wright, 1–101. Charlottesville: University of Virginia Press.

Ward, Nathaniel to John Winthrop, Jr., 24 December 1635. Winthrop Papers, 2:215–217. Boston: Massachusetts Historical Society.

Whitaker, Alexander. 1615. To my verie deere and loving Cosen M. G. Minister of the B. F. in London. In Hamor, Ralph, *A True Discourse of the Present Estate of Virginia*, 59–60. London.

Winthrop, John. 1629. *Reasons to be Considered.* Winthrop Papers, 2:138–145. Boston: Massachusetts Historical Society.

Wood, William. 1634. *New Englands Prospect.* London.

Secondary Sources

Blanton, Dennis B. 2000. Drought as a Factor in the Jamestown Colony, 1607–1612. *Historical Archaeology* 34: 74–81.

Chua, Daniel K.L., and Alexander Rehding. 2021. *Alien Listening: Voyager's Golden Record and Music on Earth.* New York: Zone Books.

Cornelius, Paul. 1965. *Languages in Seventeenth- and Early Eighteenth-Century Imaginary Voyages.* Geneva: Librairie Droz.

Craven, Wesley Frank. 1970. *The Virginia Company of London, 1606–1624.* Charlottesville: University Press of Virginia.

Cressy, David. 2006. Early Modern Space Travel and the English Man in the Moon. *American Historical Review* 111 (4): 961–982.

Ferris, Timothy. 2017. How the Voyager Golden Record Was Made. *The New Yorker: Annals of Technology*, 20 August. http://www.newyorker.com/tech/annals-of-technology/voyager-golden-record-40th-anniversary-timothy-ferris.

Goldman, William S. 2011. Spain and the Founding of Jamestown. *William and Mary Quarterly* 68 (3): 427–450.

Loomis, Edward Alleyn. 1956. Master of the Tiger. *Shakespeare Quarterly* 7 (4): 457.

Shepard, Alexandra. 2003. *Meanings of Manhood in Early Modern England*. Oxford: Oxford University Press.

Spence, Jonathan D. 1984. *The Memory Palace of Matteo Ricci*. New York: Viking.

Thomas, Keith. 2018. *In Pursuit of Civility: Manners and Civilization in Early Modern England*. Waltham, MA: Brandeis University Press.

Wood, Karenne. 2016. Prisoners of History: Pocahontas, Mary Jemison, and the Poetics of an American Myth. *Studies in American Indian Literature* 28 (1): 73–82.

Wright, Irene A. 1920. Spanish Policy toward Virginia, 1606–1612. *American Historical Review* 25 (3): 448–479.

PART II

Alternative Histories, Alternative Realities

CHAPTER 7

'I Get to Exist as a Black Person in the World': *Bridgerton* as Speculative Romance and Alternate History on Screen

Piia K. Posti

When the screen adaptation of Julia Quinn's romance series *Bridgerton* premiered in December 2020, a lot of fuss was made about the liberties taken with historical accuracy. Lists were compiled and the "mistakes" were discussed (see for example, Kickham 2021; Shanks 2022). Some fumed about the corsets, others commented on the music. But mostly, the discussion focused on the Black Queen and the unprecedented large number of Black and other non-white actors. While Shondaland, Netflix and producer-cum-screenwriter Chris Van Dusen emphasized how they were "reinventing the period drama through a color-conscious lens" (Van Dusen 2021), and actors underlined the wonderful opportunity to act in leading roles previously denied them due to their skin color,[1] some viewers and reviewers questioned the level of consciousness by which the adaptation addresses (or does not thoroughly address) race and racism. "You can't say race isn't of consequence when the world these characters inhabit

P. K. Posti (✉)
Linnaeus University, Småland, Sweden
e-mail: piia.posti@lnu.se

J. L. Hennessey (ed.), *History and Speculative Fiction*,
https://doi.org/10.1007/978-3-031-42235-5_7

was created in part through racism," states Carolyn Hinds in "'Bridgerton' Sees Race Through a Colorist Lens" (Hinds 2021).

In this chapter, I explore why the *Bridgerton* adaptation and its casting are interpreted in such a mixed and challenged way (e.g. Jean-Philippe 2020; Kini 2021). I will show that the adaptation's problems with achieving its goals of representativity and color-consciousness are related to the fact that the screenwriter and the producers have not fully considered the implications of genre. My analysis takes the genre of historical romance as one of its starting points, delves into the way race and romance are intertwined in popular Regency romance (such as Julia Quinn's *Bridgerton* novels), and shows how this creates a number of concurrent and contending interpretations of the character Simon, Duke of Hastings, when cast as Black. In order to further show why genre is so important, I discuss the adaptation's casting in relation to the British casting policy of the last decades. I consider the portrayal of some of the other Black characters in the series, such as Marina Thompson, Lady Danbury, and Will Mondrich. Also, since Van Dusen has stated that both the script and the casting were influenced by his awareness of "the historical theories of the actual Queen Charlotte's African ancestry" and how he found it "revolutionary—not just as a real, historical theory but also as the basis for the show,"[2] I pay attention to the role of history in the process and field of adaptation. Using Queen Charlotte's possible African ancestry as his starting point for the adaptation, Van Dusen also writes a kind of speculative fiction for the screen. Hence, my analysis investigates the implications of the alternate history that emerges in *Bridgerton*, a Regency world without racism, and I conclude by discussing how the adaptation may be reassessed when explored through the perspective of decoloniality.

The screen adaptation of Julia Quinn's romance series set a record for Netflix's most viewed streaming show. Season 1 was the most-watched English-language series debut of the streaming platform, with 625.49 million hours viewed within the first 28 days, only to be surpassed by season 2 and its 627.11 million hours viewed (Davies-Evitt 2022; Hipes 2022). With such immense and global impact, the adaptation also needs to be considered in the light of its transnational and transcultural status. What happens when a predominantly white genre such as popular historical romance is recast as "multihued, multi-ethnic" and broadcasted on a global scale?[3] Could it be that despite its criticized color-*un*consciousness, the show warrants a deeper analysis since it indicates how popular romance *reimagined* might in fact work as a vehicle for decoloniality?

Genre Matters: Popular Historical Romance and Racialized Scripts

Adapting Julia Quinn's Regency romance novels with Black, Asian, and minority ethnic actors (BAME actors) in leading roles challenges not only notions of historical accuracy but also the romance script itself. Popular romance is "particularly sparse in its representation of non-white characters" and the "symbolic language and imagery" of the genre is "heavily racialized" (Young 2020, 512). One aspect of this racialized symbolic language is constituted by the way the hero and the heroine are portrayed in a dichotomous relationship of dark and light characteristics. The romantic heroine's fair hair and skin are recurrently used as symbols for her innocence and inner light, a light which is meant to illuminate the darkness of the hero (Barlow and Krentz 1992, 16, 20; Young 2020, 511). The illumination of the hero is the major drive of the traditional romance plot, in which the "heroine's quest" is to encourage and "[teach] the devil to love" (Barlow and Krentz 1992, 20). Even when the heroine is portrayed as a brunette or a redhead, her fair skin is typically emphasized and linked to her purity of heart, characteristics which are then set against the dark and brooding rake who needs to be reformed and convinced of the romance ideal of domestic love and bliss.

What makes this dichotomy racialized is that the characteristics of feminine light have predominantly been ascribed to white heroines, and it is only recently that romance narratives with non-white heroines have emerged.[4] Furthermore, "English-language historical romances set in diverse locations around the globe written by non-white authors and/or starring non-white characters as the romantic leads have had an even slower growth pattern, especially from traditional outlets" (Ficke 2020, 123).[5] In a recent study of three major genres of popular fiction, *Genre Worlds: Popular Fiction and Twenty-First-Century Book Culture* (2022), Kim Wilkins, Beth Driscoll, and Lisa Fletcher further show that romance and other popular genres are marked by white privilege since "issues of racial and cultural inclusion and representation can be sidelined by white members in a genre world" (Wilkins et al. 2022, 115). Writers and publishers often operate along "industry lore," the implicit notions of, for example, the market and readers' interests, which "can lead to narrow representations of racial and ethnic groups and perpetuate unconscious bias" which often "become self-fulfilling prophecies because of how often and to whom they are articulated" (Wilkins et al. 2022, 24).[6] The "race problem" of the romance genre (Young 2020, 512) is therefore a matter both of the narrative script and of the romance industry itself.

The racialization of the romance script becomes even more visible when the hero is considered. Therefore, the choice to cast the romantic hero of the first season of *Bridgerton* as a Black Regency duke needs further analysis. Since Black men have rarely figured as romantic heroes in historical romance, such a casting decision does indeed take a radical and much needed step toward greater representativity and toward breaking the racialized script. Broadcasting a Black Regency Duke to a world-wide audience, the adaptation not only confronts the romance script, also it affects the racialized script for Black actors and Black people in general. As actor Regé-Jean Page (who plays the Duke) puts it: "I get to exist as a Black person in the world [...] It doesn't mean I'm a slave. It doesn't mean we have to focus on trauma. It just means we get to focus on Black joy and humanity" (Lenker 2020). However, by casting a Black actor as a rakish duke in a Regency romance, the Black male is inevitably reinscribed on the negative side of the familiar dichotomy of light and darkness according to the logic of the romance script. Furthermore, the logic of the romance narrative is to contain and neutralize the dark counterpart, thus privileging light and, in its trajectory, whiteness.

In the romance script, the hero, functioning as the opposite to the light and innocent heroine, is attributed characteristics such as being dark and brooding, embodying sexual danger, having "a dark and dangerous past," and being "bent upon vengeance rather than love" (Barlow and Krentz 1992, 17). He is also often portrayed as a diabolical figure, a devil (Ibid., 18–19). The Duke in *Bridgerton* certainly adheres to this script on several points. He avoids his peers and society as a result of his "dark and troubled past": an abusive father and a stigmatizing stutter. A notorious rake, he poses a danger to the heroine's virtue. And by publicly insisting that marriage and family are not in his plans, he defies the norms of society. Furthermore, the reason for this defiance is vengeance. The Duke swears on his father's deathbed that he will not continue the family line as a consequence of his father's disappointment and brutal disavowal of him when he was a boy.

This might not be seen as particularly troubling at first, since these are traits that are traditionally ascribed white heroes, and one might argue that this is an effect of a gendered script rather than of a racialized script. However, the romance dichotomy is further supported by a narrative structure that safeguards a "reproduction of whiteness" (Young 2020,

514). In the popular romance narrative, the goal of the heroine is to *conquer* the darkness surrounding the hero: "despite the hero's significantly greater social, political, and economic power, the heroine succeeds in converting him to her worldview; in effect, she transforms and domesticates him" (Young 2020, 522; see also Krentz 1992, 8; and Barlow and Krentz 1992, 19). Thus, in popular romance, whiteness is the norm. The hero's alluring dark attributes function merely as, what Toni Morrison calls, "romancing the shadow" and "playing in the dark" (qtd in Young 2020, 514), because "ultimately," the darkness of the hero is "rejected in order for 'whiteness' to prosper" (Young 2020, 514). Darkness is neutralized through the mutual love that is developed between the couple, and the (white) social order is restored.[7]

Because of the romance script, the *Bridgerton* adaptation's purported color-conscious casting policy is not as straightforward and inclusive as its director, producer, and actors seem to assume. It is true that we are now, for the very first time on a global screen, seeing a Black man in a role which has historically been denied him, in society, in popular romance fiction and in heritage film and TV series. However, the racialized codes of the romance genre simultaneously ascribe to the Black male the same kind of negative otherness which we recognize from colonial discourses and in which Black men are portrayed as sexual predators and a threat to white women and consequently white society. In the colonial context as in popular romance narratives, the "dark" male is to be controlled. Hence, in some recent romance studies, it has been argued that "popular romance contributed to the production of racial difference in support of colonial agendas" (Young 2020, 513). Paradoxically, in both discourses, there is a strong investment in the presence of darkness/blackness, but in the colonial context, the presence of actual Black men is considered threatening, and there are still very few mainstream romances in which the leading man and lover is Black. Ironically then, in the *Bridgerton* adaptation, the Black man is not only domesticated and neutralized by entering marriage and by adhering to the rules of Regency society and shouldering his duties as a Duke; he practically disappears from the series once he has fulfilled the romance script.

It could be argued that the disappearance of the Duke in the second season is simply a consequence of the fact that the popularity of the series had garnered more acting opportunities for Regé-Jean Page. Nevertheless, it cannot be denied that the *effect* is that of a strong Black character disappearing from the story once his function in the romance script has been fulfilled. The series could have recast the Duke with another Black actor (as is frequently done in other series when actors are engaged elsewhere) if Shondaland had wanted to develop the plotline of the Duke. However, in this case, the adaptation follows Quinn's plot, which in turn adheres to the rules of the romance script where the hero is operative mostly in relation to the heroine.

Since we are dealing with a TV series that has claimed a strong interest in representativity, it is important to address the effect of the Duke's disappearance from the storyline. Furthermore, it is significant for my analysis, since it reveals a *concurrence*, an instance of discursive and ideological intersections in conflict.[8] The codes and patterns of the romance narrative meet and compete with the objective of color-conscious casting, complicating the casting of BAME actors in roles such as the historical romance hero. In the romance script, the hero has a mainly functional role in providing the heroine with a worthy "opponent." This opponent is in turn, as shown above, meant to embrace the light of the heroine to ensure the stability and dominance of whiteness (Young 2020, 516): "there is a time-honored tradition of heroines sent on a quest to encounter and transform these masculine creatures of darkness" (Barlow and Krentz 1992, 19). This transformation of the hero is in turn part of the bigger quest for equal partnership: "the genre's primary drive is to imagine ways that romantic love and desire (erotic or asexual) might serve as a path to self-fulfillment and, increasingly, socio-political equality" (Kamblé et al. 2020, 3). However, according to the rules of the genre, the matter of self-fulfillment and equality predominantly concerns the heroine. In the historical romance, in particular, it is the heroine who is struggling for (fictional) recognition on her own terms, not the hero since he is already privileged. Color-conscious casting is, on the other hand, driven by demands on representativity as well as equity and acting opportunities for BAME actors. A tension is thus created between the fictional drive and the ideological drive of inclusivity in period drama casting. In the case of Shondaland's *Bridgerton*, this tension is further increased by the debate about historical accuracy. Because of Van Dusen's choice not only to cast a Black romantic hero but also to extrapolate a racially mixed Regency society from the recent claims that Queen Charlotte was Black, it is necessary to also consider how color-conscious casting and adaptations of historical narratives concurrently intersect in *Bridgerton*.

Adaptation, Racist Stereotypes, and Colorblind/ Color-Conscious Casting

Alternate history, anachronism, and unexpected casting are often used as a conscious strategy in adaptations. As Julie Sanders shows in *Adaptation and Appropriation* (2006), the point of an adaptation is usually to "make texts 'relevant' or easily comprehensible to new audiences and readerships via the processes of proximation and updating" (19). This can be done by setting the text in a different time period, by "stretching history" and "redeploy[ing it] in order to indicate those communities or individuals whose histories have not been told before, the marginalized and disenfranchised" (Sanders 2006, 140; see also Stam 2017). It can also be done through "amplificatory procedure" (adding, expanding, interpolating) with the aim of bringing the text "closer to the audience's frame of reference in temporal, geographical, or social terms" (Sanders 2006, 18, 21). For example, Hamlet may be played by a female actor in order to amplify certain aspects of the (male) character, but also in order to cast light on patriarchal structures. Other strategies of casting can draw attention to the lack of non-whites in certain genres of text and film, as with *Bridgerton*, where the unusually large number of BAME actors makes more conspicuous the white privilege of the Regency period, as well as the whitewashing in earlier period drama adaptations. Indeed, "making relevant" seems to take precedence over historical fact in the process of adaptation. Moreover, in adaptation, there is also always a consideration of the interest and pleasure of the audience, whether it is a question of relevance, identification, representativity, or simply recognition.

Consequently, historical "fact" and accuracy are not really that important in adaptation. In fact, in recent adaptation studies, historical "distortion" and anachronisms have been considered to constitute integral parts of adaptation; it has even been claimed that historiography is a form of adaptation. Defne Ersin Tutan argues in "Adaptation and History" (2017) not only that "every version of history should be regarded as a rewriting, essentially an adaptation, since the historian adapts the material [...] at hand into a pre-planned scheme to meet a certain end" (576) but also that "all historical representations are radically adaptive" (577). In her essay "Intership" (2017), Mieke Bal calls for a kind of adaptation that performs "a dialogue with the past we know was there but cannot be 'restored'" and she describes adaptation as "a productive deployment of anachronism as a figure of intertemporal thought" (181). Adaptation, and in particular the adaptation of narratives set in the distant past, is always already about much more than creating a historically accurate scene. Drawing on Tutan and Bal, I would argue that adaptation can be described as the double act of approximating the past while also establishing a dialogue between the past and the present.

With its emphasis on "making relevant," it is hardly surprising that adaptation has also been one of the most significant areas in which contemporary interest in representativity and the promotion of diversity have managed to gain traction. In her 2020 article, with the telling title "Casting for the Public Good," Christine Geraghty "examines changes in casting practices which have begun to put black, Asian, and minority ethnic actors more regularly on British screens and in more significant parts" (168). The major driving forces behind the changes in casting policies have been equality and justice, "to improve the employment position" of BAME actors (169). According to Geraghty, this call for equity has been addressed by using two strategies: (1) adapting narratives from a "wider range of experience," as stories by and about the lives of BAME people would be performed by BAME actors and (2) by making "more roles ... available for BAME actors," by adopting a policy of colorblind casting (Geraghty 2020, 169). Operating as a meritocratic policy, colorblind casting assumes that any actor may be chosen for any role, irrespective of race and ethnicity. However, despite the fact that British theatre has used colorblind casting for quite some time now, it is only during the last decade that film and television have started to practice colorblind casting.[9] Geraghty argues that this lag has to do with the higher expectations of verisimilitude and historical accuracy for film and television; the medium as such carries with it a notion of "truth" (Geraghty 2020, 175). Theatre audiences however are expected to understand the "conventions" of the stage and "know that they should make a distinction between actor and character and that, in this context and on this stage, ethnic origin and colour of skin are not significant in creating meaning" (Geraghty 2020, 170).

It has been argued, however, that these conventions rather than promote diversity might just as easily emphasize skin color while negating the experiences of people of color (Geraghty 2020, 172). Ayanna Thompson, Christine Geraghty, and L. Monique Pittman have all provided examples of when colorblind casting assumes a double standard of both seeing and not seeing race and ethnicity. Thompson argues that "nontraditional casting can actually replicate racist stereotypes *because* we have not addressed the unstable semiotics of race" (Thompson 2011, 77). Referencing Thompson, Geraghty maintains that: "On the one hand, audiences are expected not to take ethnicity and skin colour as semiotically significant while, on the other, they are meant to recognize that the cast as a whole represents a society marked by multicultural diversity" (Geraghty 2020,171). Similarly, Pittman contends that "the colourblind casting strategy often fails to correct problems of multicultural representation and does not adequately account for the way in which an actor's skin colour sets in motion unintended racialized meanings" (Pittman 2017, 182).

In the *Bridgerton* adaptation, racialized meanings occur several times in scenes with the Black characters. We have already seen how the Duke is associated with an image of darkness that resonates with colonial sentiments, and I will show presently how the Duke's working-class friend the boxer Mondrich is linked with slavery and ends up in a scheme of being sold for the benefit of a white man and his family's social future among the ton. Throughout the series the casting strategy creates a concurrence of colorblindness and color-consciousness that problematizes its claims to representativity and diversity. Thompson's cautioning that "nontraditional casting can … replicate racist stereotypes" definitely holds for several of the Black roles in *Bridgerton*.

Lady Danbury and boxer Will Mondrich both constitute examples of such stereotypical replication. A long-awaited powerful Black woman in a period drama, Shondaland's Lady Danbury has been applauded for finally breaking the racial glass ceiling of heritage film and TV series (alongside the character of Queen Charlotte). However, as Rachel Bass has argued, "one must look beyond the status given to her character and instead at the ideology she is expressing" (Bass 2023). In her analysis, the power of Lady Danbury, as well as Queen Charlotte, is very much directed at "keep[ing] whiteness as the focal point" and their function in the story is to convey "a loyal maternalism reminiscent of the minstrel mammy" (Bass 2023). I would argue that Lady Danbury not only replicates the minstrel mammy but also other similar stereotypes known from film history. Her lack of a background story—how did she end up such a wealthy widow and with such a high position in society?—and the way she supports the white characters makes her function like a female variety of the "ebony saint," an upscaled and more positive version of the "Magical Negro" ("MN") familiar from both fiction and film, and the similar "Magical Asian" character, whose purpose in the narrative are "to further the white character's journey."[10] The trope of the "MN" has a long history in American cinema and it refers to

> a stock character that often appears as a lower class, uneducated black person who possesses supernatural or magical powers. These powers are used to save and transform disheveled, uncultured, lost, or broken whites (almost exclusively white men) into competent, successful, and content people within the context of the American myth of redemption and salvation. (Hughey 2009, 544)[11]

The "ebony saint" trope, Matthew W. Hughey asserts, appeared in conjunction with the emergence of "an assimilationist agenda of sanitized African American characters [...] during the late 1950s and early 1960s as a response to demands for more 'positive' characters" (Hughey 2009, 545). This trope is associated with notions of integration and assimilation, providing a non-threatening Black character on screen who enjoys some modicum of equality in white society by playing by its rules.

Despite the fact that Lady Danbury has an even more dominant role in the adaptation than in the novels, and one which may at first glance look like a conscious move to give a Black actor a prominent and strong role, the character's importance derives mostly from the way she supports the white characters, the power structures, and the ideals of the society. She is the one who glosses over the racist past of their society when emphasizing the benefits that the Black people of the ton have accessed through the marriage of the King to "one of us": "look at everything it is doing for us, allowing us to become" (S. 1, Ep. 4). She sympathizes with the heroine of the romance, Daphne, and her struggles with her overbearing brother. She encourages and facilitates the Duke's courtship of Daphne, their marriage by special license, and she safeguards their reconciliation when stopping the rest of the ton from dancing and sends them home at the final ball at the end of season one. The ball scene is significant since this is meant to be the last time the Duke and Daphne will attend to their social duties as a couple. After the ball, they have planned to remain married but lead separate lives. However, by stopping the ton from dancing, Lady Danbury provides the couple with the private romantic moment they need to reconcile and reach the final stage of domestic bliss. This might be interpreted as her only helping the Duke, whom she has taken under her wing since his childhood (like a benevolent "mammy"), but considering the fact that Daphne is the lead character of the first novel and the first season of the adaptation and that the entire narrative is about the (white) Bridgerton family, her support is ultimately a support of white heroes and heroines.

Moreover, like previous "ebony saints" who enjoy some privilege since they are assimilating to white society, Lady Danbury can function as a supporting character to the entire Bridgerton family due to her position as advisor to the queen with access to the core of the power (mad King George). In season two, she has a hand in choosing Edwina as the Diamond of the season, and she is the one Queen Charlotte turns to with demands to "repair today's situation" and sort out the mess with the Viscount's wedding (S. 2, Ep. 6). Even though Lady Danbury is neither poor nor uneducated, she is arguably reminiscent of the facilitating "MN" and the assimilating "ebony saint." Thus, the example of Lady Danbury provokes further questions about the level of "consciousness" by which the series was cast and reinforces the negative statements about the series' "surface-level diversity" (Komonibo and Newman-Bremang 2020).

Another character who displays several characteristics of the two tropes is the boxer William Mondrich. The boxer and his family do not appear in the novels, and being an added element in the adaptation, this character becomes even more interesting to consider in relation to the series' color-conscious assertions. Some viewers have been puzzled, to say the least, by this character's purpose in the first season. Carolyn Hinds points out in the *Observer* that he is one of the only two truly dark-skinned Black actors in the series and that it is quite telling that he serves "as Simon's unpaid therapist" (Hinds 2021). This character functions as a foil to the Duke in the first season, as he offers a space (the boxing ring) where the polished Duke can relax, exert himself physically, and receive counselling on his journey to reconcile with his past and to develop into a proper romantic hero, the reformed rake.[12] In the beginning, Mondrich mostly plays the part of the uneducated and lower-class character who has an extraordinary ability of strength and fighting stamina, being a boxing champion. As the series develops, however, Mondrich further approximates the "MN" as an aid to and supporter of the white characters. Mondrich is persuaded to participate in a betting scheme to help Lord Featherington pay his debts, which in turn would allow his wife and daughters to buy new dresses (furthering their chances on the marriage market) and allow Philippa Featherington her dowry and wedding (S. 1, Ep. 8). Here, Mondrich becomes the linchpin for the fate of the (white) Featherington family.

In the narrative arc of Colin Bridgerton's revelation of the new Lord Featherington's embezzlement scheme, Mondrich plays a supportive role in ways that are once again reminiscent of the "MN" trope. First, Mondrich seems to have a unique ability to detect dishonesty, and he draws attention to the new Lord's dubious character (S. 2, Ep. 6). Second, he retreats when Colin stands up for Lord Featherington as part of his ploy to expose the Lord's shady dealings (S. 2, Ep. 8), thus allowing the white man to be the hero proper and save the entire ton from possible financial ruin. And third, Mondrich is rewarded by the white hero for stepping back and granting him the role as savior (according to the narrative pattern in which the trope figures): Colin brings Mondrich the much-needed patronage of his newly opened club.

In this way, both Lady Danbury and Mondrich display several characteristics of the "MN" trope and the "ebony saint," which in turn troubles the claims made about the series' color-consciousness. Tropes like these manifest what Hughey terms "cinethetic racism," a version of a "new racism" as it "reinforces the meaning of white people as moral and pure characters while also delineating how powerful, divine, and/or magic-wielding black characters may interact with whites and the mainstream." The insidiousness of "new racism" is that it "supports the social order while seemingly challenging the racial inequality constitutive of that order" (Hughey 2009, 544).

As a result, the romance script and the casting of *Bridgerton* intersect in a way that reinforces a "new racism" while simultaneously providing a global audience with an unforeseen level of diversity and representation in historical romance and heritage film and TV series. While opting for an alternative history in which racism has ceased through a racially mixed royal marriage, the adaptation nonetheless, in the words of Shaun Armstead, "traffics in historical fantasy seeking to emulate the liberal politics of our present. … [I]nstead of breaking down historical racial and gender formations that continue to dehumanize Black women and men today, *Bridgerton* reinforces them" (Armstead 2021).[13] In fact, the way the adaptation attempts to counteract the racism of the period and promote diversity in the present is quite similar to the way white intellectuals in the 1700s and 1800s transferred the suffering of African slaves to their own moral suffering within the system of slavery. Black experience was exploited within the emancipation movement by reformed whites and white abolitionists. While this was sometimes done to evoke empathy for the enslaved, by forging a common experience, it was still a discourse of suffering that centered upon whiteness and white sentiment (Wood 2002, 23–86). As Marcus Wood has maintained, this transference hinged upon dehumanizing Blacks (Wood 2002, *et passim*). It seems that both then and today, breaking down racial formations involves acts of dehumanizing Black people. Such dehumanizing also occurs in *Bridgerton*. Despite its reimagination of a Recency period in which Black and white people have equal opportunity, there are two scenes in which the series' dehumanizing tendencies thoroughly surface.

These two scenes make actor Regé-Jean Page's jubilant words about not being portrayed as a slave and dehumanized seem rather hasty. I am referring here to the scene in which Marina Thompson, a cousin of the Featheringtons, is offered to an old bachelor and to the scene in which Mondrich is asked by Lord Featherington to forfeit a boxing match. These scenes illustrate how the casting of BAME actors in *Bridgerton* tends to support a racist script rather than dismantle it.

Marina Thompson is interesting in several ways. Firstly, she only appears as a character with a developed narrative arc in the screen adaptation; Marina is wholly a creation of Van Dusen and Shondaland. Secondly, viewers have commented on the fact that "Marina's character is both relentlessly commodified and hyper-sexualised as a stereotype of the 'tragic mulatta'" (Garden 2021). Hence, the character is not seen by the audience through a colorblind lens. Rather, since this is a case in which the "role overlaps with cultural stereotypes associated with the ethnicity of the actor in question" (Pittman 2017, 182), Marina is interpreted as an example of stereotyped Black characters which in turn makes the color-conscious casting ambition quite ambiguous. Thirdly, there is one scene in particular in which viewers are clearly meant to pay attention to the actor's skin color. When Mrs. Featherington discovers that Marina is with child, there is a rush to secure a marriage for her to avoid a scandal. An elderly and unsympathetic bachelor is invited to the house at which point he demonstrably inspects Marina's teeth and body in a way that revokes the denigrating inspections of Africans on the slave market (S. 1, Ep. 4).

With the example of Marina, we are yet again given an illustration of the way popular historical romance "can serve to erase problematic issues and reinforce white fantasies of the time" (Hernandez-Knight 2021, 1) and why casting BAME actors in historical period drama can cause ambivalence when the adaptation has not thoroughly addressed the racism of the period and the racist legacy within the historical romance genre. In fact, the sexualized Black woman at the mercy of the slave owner and images of Black people as chattel are such established cultural and literary tropes that "for black women readers, romantic desire for white men cannot be easily divorced from the historical legacy of abuse and dehumanization that black women have experienced at the hands of white men and the institutions they control" (Young 2020, 523).[14] We may therefore ask what purpose this scene is meant to serve in the *Bridgerton* adaptation. Is it there to provoke awareness of racist society (then and now)? Or is it simply another example of what Pittman describes as "unintended racialized meanings" (182)? Moreover, are those meanings really so unintended?

The inspection scene creates further ambivalence in that it also resonates with the equation made by European women writers already in the seventeenth century between slavery and the marriage market. French literary examples and strategies have been traced by Karen Offen in "How (and Why) the Analogy of Marriage with Slavery Provided the Springboard for Women's Rights Demands in France, 1640–1848" (2007). In a British context Moira Ferguson noted in the early 1990s that even Jane Austen, who has repeatedly been read as problematically silent on British imperialism and the slave trade, "connects the Caribbean plantation system and its master-slave relationships to tyrannical gender relations at home and abroad" (Ferguson 1991, 130). The "comparison of the young woman on the verge of marriage with the slave on the auction block" was already established by the early eighteenth century by Mary Astell and Judith Drake in *An Essay upon the Female Sex* and Mary Astell in *Some Reflections upon Marriage*, and it was further employed by writers such as Jane Austen in *Emma* and George Eliot in *Daniel Deronda* (Wood 2002, 313–314).

There is little doubt that the scene in *Bridgerton* constitutes a contemporary version of the trope. However, since this trope is mainly associated with the struggle of *white* women (both historically and in the literary narratives), it can be interpreted as yet another white fantasy that has been reiterated within literary romances and travelled to popular historical romance. Notably, this possibility of interpretive recognition of a racist legacy is not duplicated *within* the world of the adaptation. Whereas the audience is invited to notice skin color to access such interpretive layers in certain scenes as well as partake of the fictional representation of a diverse, inclusive, and colorblind Regency Britain, there is hardly any recognition of a racist past in the adaptation itself. The shift that apparently has taken place from a racist to a harmonic and fully diverse society in *Bridgerton*'s alternate history seems to have caused selective amnesia in the white characters. It is mainly the Black characters who voice any awareness of racism.

The white characters act as if instructed by a colorblind script. This becomes glaringly apparent when we consider Eloise Bridgerton's critique of marriage, her highly verbalized awareness of social injustice against women, and her brother Benedict's struggle against classist ideas of what kind of occupation is befitting a gentleman's second son. Why does neither display any memory of other forms of injustice such as their society's recent racism?

The one exception is Lord Featherington who makes a blunt remark about boxer Will Mondrich's past and about his father being an American slave. This takes place in the scene in which Mondrich refuses to throw his next fight by maintaining "My honor is not for sale" (S.1, Ep.7). Here Mondrich's insistence that he does not wish to sell his honor carries double meaning because of Featherington calling him out as the son of a slave. Yet this can easily go unnoticed as being a racist remark since not all viewers might know that Featherington's mentioning of "Dunmore's regiment" in the scene is a reference to an actual British regiment of escaped American slaves who were offered freedom if they fought on the British side in the American revolution (Quarles 1996, 19–32). Since this remark is brief and easy to miss—it rests wholly on the mentioning of the regiment—the scene functions rather as a demonstration of the level of Featherington's moral corruption than to truly "address" the "semiotics of race," to repeat what Ayanna Thompson argues is often lacking in cases of "nontraditional casting" (Thompson 2011, 77). The reference to Dunmore's regiment shows that the audience is still meant to notice that Mondrich is Black and pick up on the racism, while also going along with the alternate history of a diverse and non-racist Regency society, which rests upon the speculative notion that love conquers (and erases) all.[15]

It Takes a (Black) Queen: Speculative Romance and Decoloniality

> I understand that you believe such subjects as love and devotion, affection and attachment, you find it all trite and frivolous. But have you any idea, those very things are precisely what have allowed a new day to begin to dawn in this society. Look at our Queen. Look at our King. Look at their marriage. … We were two separate societies, divided by colour, until a king fell in love with one of us. Love, Your Grace, conquers all. (S 1. Ep. 4)

These lines, spoken by Lady Dunbar to the Duke, constitute the only explicit commentary on the series' alternate history and explanation of its diversity. In that regard, they provide the audience with an important cue for reading the *Bridgerton* adaptation as speculative fiction. They also reveal that it is a speculative fiction that operates from two prerequisites: the alternate history of a Regency Britain ruled by a Black Queen Charlotte and the logics of romance in which love is portrayed as a powerful force that can right any wrong. Note once more the importance of the romance genre for understanding the alternate world of this adaptation. In fact, romance fiction can in itself be seen as a form of speculative fiction by its insistence upon love as the universal means for resolving anything from gender inequality, family conflict, enmity, classism, ageism, to racism. As Catherine M. Roach argues, and in line with previous claims made by romance scholars like Janice Radway, Tania Modleski, and Ann Barr Snidow, "romance stories are a creative respite for women, an imaginative play space to roll around in the fields of fantasy with sister readers of the genre, all the while affirming the reality of love as a force that can work good in the world" (Roach 2016, 13).

Speculative fiction takes various forms (fantasy, utopia, dystopia, science fiction) and can be defined in many ways, but at its core we find a "what if?" (Thomas 2013, *et passim*). In her short essay "Writing *Oryx and Crake*," Margaret Atwood states that speculative fiction "begins with a *what if* and then sets forth its axioms" (Atwood 2005, 285). In popular historical romance, that *what if* is often formulated as: What if the heroine could enjoy heterosexual love and all the riches and privileges of aristocratic society while also managing to bypass the constraints of patriarchy? As stated in the introduction to this volume, speculative fiction is also a way of making the familiar unfamiliar and the unfamiliar familiar: "it invents nothing we haven't already invented or started to invent" (Atwood 2005, 285). The *Bridgerton* adaptation makes the familiar world of period screen drama unfamiliar by challenging its white gatekeeping and whitewashing, while simultaneously familiarizing the audience with an alternate and unfamiliar non-racist Regency Britain.

Speculative fiction is a way of dreaming that which has not yet been realized. It creates a fantasy about a different way of living and being, with a power to make us question our actual lives, beliefs, and existence. As activist scholar Stephen Duncombe argues:

> Dreams are powerful. They are repositories of our desire. They animate the entertainment industry and drive consumption. They can blind people to reality and provide cover for political horror. But they can also inspire us to imagine that things could be radically different than they are today, and then believe we can progress toward that imaginary world. (Duncombe 2007, 182)

The phrases imagining the "radically different"' and "progress toward that imaginary world" can also describe the method and objective of decoloniality. What speculative fiction and decoloniality have in common is the drive to challenge our thinking, to question what we know, "to disengage from the illusion and focus on the puppeteer behind the scenes, who is regulating the terms of the conversation" and to "alter… the principles and assumptions of knowledge creation, transformation, and dissemination" (Mignolo 2018, 144–145). Simply put, both aim to tell a different (hi)story in a different way.

Defining decoloniality as an analytics and praxis that aims to "delink" from "the logic of coloniality," Walter D. Mignolo argues that postcolonialism and decolonialism have not succeeded in achieving such a delinking: "The changes were in the content not in the terms of the conversation" (Mignolo 2018, 125 & 124). The change that needs be made he describes as follows:

> By conceiving of coloniality … as a complex structure of management and control, one grasps that it is the "underlying structure" of Western civilization and of Eurocentrism and that fully understanding how it works is a necessary condition for delinking from coloniality. Eurocentrism is not a geographical issue, but an epistemic and aesthesic one (e.g., control of knowledge and subjectivities). In order to do so, it is necessary to think and act (doing, praxis) decolonially, both in the analysis of the colonial matrix of power—delinking subjectively and programmatically from it—and by engaging with projects and organizations that run parallel and in the same direction. (Mignolo 2018, 125)[16]

One form of delinking or praxis of decoloniality would be to challenge and reimagine "universal fictions" (Mignolo 2018, 187). Therefore, imagining alternate histories and creating speculative fictions, such as popular historical romance, can be delinking since such histories and fictions often engage with and challenge the universal fictions at work in our reality. In *The Glass Bead Game*, for instance, Hermann Hesse speculates in the "what if there were a world where aesthetics and philosophy were in focus and technology and economy were reduced to a bare minimum?" In this narrative it is the universal fiction of Western capitalism that is challenged and reimagined by Hesse's creation of a future world in which technology and economy are devalued and intellectual and spiritual pursuits are glorified.[17] In Shondaland's *Bridgerton*, the universal fiction of white supremacy is challenged and reimagined through the lenses of diversity and racial equality. These two narratives certainly do not address the same universal fiction, and *The Glass Bead Game,* with its aesthetic principles, is arguably complying with colonial epistemic and aesthetic ideals. Nevertheless, they do share the speculative form as a method to question universal fictions and to dream an alternate reality.

Even though *Bridgerton,* as shown in my analysis, does not resolve the universal fiction of white supremacy underlying Britain's colonial legacy and the romance script, it could, from the perspective of decoloniality, still be argued that it engages in and has caused global engagement in analytic and "praxical" work of a decolonial kind. As Mignolo maintains, there are reimaginings that "have a decolonial import and dimension, though they might not define themselves as decolonial, and are remodeling knowledge, being, and communal relations" (Mignolo 2018, 126). Reimaginings of racism and sexism, for example, can have such decolonial import (Ibid.). Hence, if we shift our perspective from the *content* of the conversation to the *terms* of the conversation, we might discover that *Bridgerton*, despite its entanglement in a racist script, also constitutes a change of the terms. And this may indeed explain why the adaptation has been received with both praise and criticism.

Because *Bridgerton* is a phenomenon that includes much more than the actual adaptation. It is a transnational and global event that has inspired numerous articles and analyses in popular media, countless conversations on social media platforms, and disputes in forums dedicated to literary romance (Jane Austen) and other popular adaptations of Regency romance (*Sanditon*) (Prescott 2021). *Bridgerton* has been called out on its inherent racism, and in these conversations, the racism among fans of heritage drama has been revealed. As a study by Amanda-Rae Prescott reveals, "*Bridgerton*'s premiere … changed the media environment and also revealed new divisions in Austen and Austen-adjacent spaces. The promo photos of [the Black actors] quickly resulted in the 'historical accuracy means no BIPOC actors in traditionally white roles' racists rushing to trash the series" (Prescott 2021, 5). The phenomenon of *Bridgerton* has forced media, fandom, and scholars to engage in a conversation about the unstable semiotics of race, across the boundaries of their respective spheres.

Bridgerton is a testimony of injudicious casting, but it is also an example of decolonial praxis. According to Van Dusen, *Bridgerton* was created through the work of listening, dialogue, and a demand to do more:

> My goal of reinventing the period drama through a color-conscious lens was taking shape. But then several members of my brilliant cast reached out suggesting I do more. That's when one of the most unexpected and satisfying collaborations of my career happened.
>
> What followed was one of the most poignant and transformative days I had during the making of this series. Together with every single actor of color on the show in one room, I was able to listen to everything everyone had to say over a long afternoon of tea and other English goodness. My job was to simply sit, listen and learn. It was emotional, powerful and completely necessary.
>
> Many of those in that room felt the show could go further in terms of its exploration of race. The show, they agreed, was already so beautifully eloquent when looking at things like class, gender and sexuality. But couldn't there also be an acknowledgment of color onscreen?
>
> The question humbled me. They were right. We could do even more to turn the genre on its head and dig even deeper into the stories of the characters the show aimed to include. So the things my cast talked to me about that day found their way into the scripts. Into the characters' backstories. Into the world itself. (Van Dusen 2021)

Van Dusen's narrative includes phrases that could indeed cause skepticism and even frustration among postcolonial critics and BIPOCs. What kind of allyship is described here? Why the emphasis on tea and "English goodness"? However, we need also keep in mind that decolonialty is a work that can take many (imperfect) guises:

> There is no master plan and no privileged actors for decoloniality. There are, certainly, scales in the intensity of colonial wounds. Decoloniality is a multifaceted global enterprise in the hands of the people who act and organize themselves/ourselves as decolonial thinkers, actors, and doers. If coloniality is all over, decolonial praxis shall be over as well. Consequently, no experience of privilege could be claimed in the complexity of global decoloniality. (Mignolo 2018, 125)

If we shift our attention from what is being said to what is being done in Van Dusen's description, we can see that *Bridgerton* was created through the decolonial work of dialogue between representatives of groups that do not normally converse upon matters of race and racism in heritage film. Moreover, as Prescott has shown, in popular media and on social media platforms, new conversations and analyses are taking place that did not take place before the impact of the reimagined world of Shondaland's *Bridgerton*. All contribute to "changing the terms of the conversation" which Mignolo argues is pivotal to achieve decoloniality.

To conclude then, the reception of *Bridgerton* shows that as speculative romance and alternate history it can indeed participate and even provide important "work" in "delinking" from coloniality. For example, one effect of the casting policies of *Bridgerton* is that "BIPOC fans have pushed for *Sanditon* to expand Crystal Clarke's role as Georgiana and diversify the cast and crew" (Prescott 2021, "Abstract"). (Georgiana is the adaptation's version of the "mulatta" heiress, Miss Lambe, in Jane Austen's unfinished novel *Sanditon*.) Thus, *Bridgerton* contributes to familiarizing audiences on a global scale with the presence of Black people in Regency Britain (which is historically accurate), to foster a broader interest in learning about the lives of Black people of that period, to recognize the presence of Black characters (however minor) in the work by canonized writers like Jane Austen, to discover the absence of Black characters in contemporary Regency romance, and to reflect upon casting and matters of representativity. *Bridgerton* has also contributed to further discussions of "white privilege" in many arenas and on many levels (romance, casting, fandom).

But as my analysis shows, that decolonial work and "delinking" takes place in a concurrence of conflicting and competing scripts, conventions, adaptive choices, and casting policies that affect and inform the narrative, causing equally conflicting impressions and interpretations of the TV series. This in turn demands methods of analysis that address the complexity of the concurrences. In this chapter, I have addressed the complexity by considering not only the "text" of the adaptation but also how characters and scenes take on different meanings depending on how we understand the effect of genre (historical romance, speculative fiction), casting choices and policies, demands on historical accuracy, matters of adaption and adaptation theory, cinethetic racism, fandom, and popular media. I have shown that even popular phenomena such as heritage film and TV series can contribute to decolonial work, but also that the work of scholars within Romance fiction studies, adaptation studies, and studies of fandom can contribute to further the understanding and choices of writers, screenwriters, and producers. However "imperfect" the result of the casting choices and explorations of the Black characters, *Bridgerton* shows us that the analysis should include more that the adaptation "text" if we are to grasp what impact this global phenomenon may have on the colonial legacy.

Notes

1. Golda Rosheuvel (Queen Charlotte) celebrates the series for its inclusivity: "I knew that it was something that I was hoping to see. That I have *always* been hoping to see: that inclusion, that diversity, pushing the boundaries so Black and brown artists can be celebrated in fabulous clothes and fabulous wigs" (Burack 2022).
2. Van Dusen 2021. Independent historian Mario de Valdes y Cocom is most often referred to in the newspaper articles that have spread to a larger audience the theory of Britain's first Black royalty. See for instance, Stuart Jeffries, "Was this Britain's first black queen?", *The Guardian*, 2009.
3. I have borrowed the quoted words from Van Dusen 2021.
4. More inclusive romance fiction began to emerge in the twenty-first century, according to Young 2020, 512.
5. It should be noted that Julia Quinn's *Bridgerton* series does not include non-white characters.
6. See also Wilkins et al. who argue that genres "can exclude. They may do this by reproducing enduring structural inequalities, on the basis of gender or race, for example, limiting some people's ability to participate or lead" (Wilkins et al. 2022, 113).
7. Paradoxically, in the *Bridgerton* adaptation, (white) order is assisted further by the female Black characters. For an astute analysis of how "the show creators center white womanhood as the most desirable, beautiful and marriageable, and uses Black women as her most stalwart advocates," see Rachel Bass, "She Wins: Here's to Powerful Black Women Leaders on Screens," 2023.
8. *Concurrences* is to be understood as both a methodological tool and a theoretical perspective. It was established to counter challenges within postcolonial research and has been explored in different forms by a growing research group at the Linnaeus University in Sweden. Gunlög Fur, to whom this book is dedicated, is among the founding members of the group, including Hans Hägerdal, Peter Forsgren, Maria Olaussen, Margareta Petersson, and myself. The epistemological objective of identifying and analyzing *concurrences* in a postcolonial and decolonial context is

to provide a framework and method in order to "capture" and "encompass" the "topics and challenges of difference, entanglement and complicity that postcolonial studies has brought to light and grappled with in the last decades." As a conceptual framework, *concurrence* "draws attention to the contestations over epistemic entitlement, competing (and sometimes conflicting) narratives of (post)colonial encounters and experiences, and territorial claims, with which studies with a global perspective invariably must grapple" (Posti 2014, 1337–1338.). It also "offers a way of thinking about similarity and difference together, without necessarily privileging the priority of one over the other and without assuming the parameters of relationality in advance. To look for concurrences is not to assume either full equivalence across systems or the inferiority of one to another" (Brydon, Forsgren and Fur 2017, 3). For a more extensive discussion of the concept as well as studies where the concept and perspective are used, see Brydon et al. 2017.

9. See Pittman for a discussion of the *The BBC's Diversity Strategy 2011–15*. See Geraghty for the BBC Guidelines for diversity and the way the British Film Institute ensures more diverse casting. Both Pittman and Geraghty compare the longer use of color-blind casting in British theatre with its more recent use in film and television.
10. Carrie S. quoted in Young 2020, 526. The quotation is from an online review by Carrie S. of Mary Balogh's novel *Someone to Love*. The review is posted on *Smart Bitches Trashy Books,* and I have included the quote here and its source history with the purpose of demonstrating that there is today public awareness of such racist stereotyping of Black and Asian characters.
11. A few examples of "MN" characters are John Coffey (played by Michael Green Duncan) in *The Green Mile* (1999); Candelaria (Veryl Jones) in *The Punisher* (2004); and The Oracle (Gloria Foster and Mary Alice) in *The Matrix* films (1999–2003).
12. Even though the Duke is played by a Black actor, the character as such, a Regency rake, is highly associated with white privilege within the historical romance genre (see Young 2020, 513–520 and my previous discussion). Hence, the insertion in the adaptation of a new Black character positions that character in a complex relationship within which the character (Mondrich) nonetheless functions as a kind of "MN" or "ebony saint" supporting the white (although cast as a Black) hero/reformed rake of the narrative. We may recall Morrison's notion of "playing in the dark," which is meant to draw attention to the fact that it is never a question of embracing Black identity but of creating an "Africanist presence" to set off the heroism and ideals of white society (Young 2020, 514).
13. There is quite an irony to this emphasis on the marriage between a Black Queen and white King as the solution to ending racism when considered

in the contemporary context of what has unfolded in the media about the British royals after Meghan Markle and Prince Harry were interviewed by Oprah Winfrey in 2021.

14. For an extensive study of how Black women writers have contested and negotiated the literary legacy of dehumanization and oversexualization of Black women, see Patton 2000.
15. Once again, we see how the strategies of color-conscious casting concur with the racialized romance script. As Sarah H. Ficke has pointed out, "traditional historical romance stories rely upon limited, sanitized settings or the erasure of dehumanizing political and economic systems" (Ficke 2020, 124).
16. In the quoted passage, Mignolo uses the less common term "aesthesic" to remind us of the fact that the universal fictions of coloniality also affect the sensory level: "Universal fictions operate on our sensibilities; they have an *aesthesic* power, affecting our senses, driving our emotions and desires" (Mignolo 2018, 187).
17. Hesse "bemoaned the pressures generated by life in modern capitalist society and the instrumental and often callous behaviour which threatened to engulf the world. In response he espoused the pursuit of self-understanding through grasping the spiritual insights offered to us in tragedy and the illusory nature of prestige and wealth. Occasionally he expressed sympathy with the ideal of socialism and even admired Marx, considering his critique of capitalism 'essentially incontrovertible'" (Wilde 1999, 87).

References

Armstead, Shaun. 2021. Blackness, Dehumanized: A Black Feminist Analysis of "Bridgerton". AAIHS African American Intellectual History Society. February 2. Accessed January 30, 2023. https://www.aaihs.org/blackness-dehumanized-a-black-feminist-analysis-of-bridgerton/.

Atwood, Margaret. 2005. Writing *Oryx and Crake*. In *Writing with Intent. Essays, Reviews, Personal Prose: 1983–2005*, 284–286. New York: Carroll and Graf Publishers.

Bal, Mieke. 2017. Intership: Anachronism between Loyalty and the Case. In *The Oxford Handbook of Adaptation Studies*, ed. Thomas Leitch, 179–196. New York: Oxford University Press.

Barlow, Linda, and Jayne Ann Krentz. 1992. Beneath the Surface: The Hidden Codes of Romance. In *Dangerous Men and Adventurous Women: Romance Writers on the Appeal of the Romance*, ed. Jayne Ann Krentz, 15–30. Philadelphia: University of Pennsylvania Press.

Bass, Rachel. 2023. She Wins: Here's to Powerful Black Women Leaders on Screens. *Ms. Magazine*, January 7. Accessed January 30, 2023. https://msmagazine.com/2023/01/07/golden-globes-viola-davis-black-women-bridgerton/.

Bridgerton. 2020. *Season 1. Shondaland.* USA: Netflix.

———. 2022. *Season 2. Shondaland.* USA: Netflix.

Brydon, Diana, Peter Forsgren, and Gunlög Fur, eds. 2017. *Concurrent Imaginaries, Postcolonial Worlds: Toward Revised Histories.* Leiden: Brill.

Burack, Emily. 2022. Golda Rosheuvel Thinks *Bridgerton's* Queen Charlotte Sees Herself in Edwina. *Town & Country*, March 31. Accessed August 24, 2022. https://www.townandcountrymag.com/leisure/arts-and-culture/a39529646/golda-rosheuvel-queen-bridgerton-season-2-interview/.

Davies-Evitt, Dora. 2022. *Bridgerton* Season 2 Breaks its Own Record as the Most-Watched English-Language Series on Netflix. *Tatler*, April 21. Accessed May 22, 2022. https://www.tatler.com/article/bridgerton-season-2-breaks-its-own-record-as-the-most-watched-english-language-series-on-netflix.

Duncombe, Stephen. 2007. Dreampolitik. In *Dream: Re-Imagining Progressive Politics in an Age of Fantasy*, 176–183. New York: The New Press.

Ferguson, Moira. 1991. *Mansfield Park*: Slavery, Colonialism and Gender. *Oxford Literary Review* 13 (1–2): 118–139.

Ficke, Sarah H. 2020. The Historical Romance. In *The Routledge Research Companion to Popular Romance Fiction*, ed. Jayashree Kamblé, Eric Murphy Selinger, and Hsu-Ming Teo, 118–140. London: Routledge.

Garden, Alison. 2021. Bridgerton and Normal People Expose Romance's Colonial Hangover. *Aljazeera*, April 20. Accessed September 8, 2022. https://www.aljazeera.com/opinions/2021/4/20/bridgerton-and-normal-people-expose-romances-colonial-hangover.

Geraghty, Christine. 2020. Casting for the Public Good: BAME Casting in British Film and Television in the 2010s. *Adaptation* 14 (2): 168–186.

Hernandez-Knight, Bianca. 2021. Race and Racism in Austen Spaces: Jane Austen and Regency Romance's Racist Legacy. *ABO: Interactive Journal for Women in the Arts 1640-1830* 11 (2): 1–16. Article 12.

Hinds, Carolyn. 2021. "Bridgerton" Sees Race Through a Colorist Lens. *Observer*, January 1. Accessed March 14, 2022. https://observer.com/2021/01/bridgerton-sees-race-through-a-colorist-lens/.

Hipes, Patrick. 2022. Bridgerton's Season 2 Edges Season 1 To Become Netflix's Most-Watched English-Language Series Debut. *Deadline*, April 19. Accessed May 20, 2022. https://deadline.com/2022/04/bridgerton-season-2-netflix-record-views-1235005908/.

Hughey, Matthew W. 2009. Cinethetic Racism: White Redemption and Black Stereotypes in 'Magical Negro' Films. *Social Problems* 56 (3): 543–577.

Jean-Philippe, McKenzie. 2020. *Bridgerton* Doesn't Need to Elaborate on Its Inclusion of Black Characters. *Oprah Daily*, December 29. Accessed August 15, 2022. https://www.oprahdaily.com/entertainment/tv-movies/a35083112/bridgerton-race-historical-accuracy/.

Jeffries, Stuart. 2009. Was this Britain's First Black Queen? *The Guardian*, March 12. Accessed September 8, 2022. https://www.theguardian.com/world/2009/mar/12/race-monarchy.

Kamblé, Jayashree, Eric Murphy Selinger, and Hsu-Ming Teo. 2020. Introduction. In *The Routledge Research Companion to Popular Romance Fiction*, ed. Jayshree Kamblé et al., 1–23. London: Routledge.

Kickham, Dylan. 2021. 9 Historical Inaccuracies In "Bridgerton" Season 1 You Won't Be Able To Unsee. *Elite Daily*, January 29. Accessed September 8, 2022. https://www.elitedaily.com/entertainment/9-historical-inaccuracies-in-bridgerton-season-1-you-wont-be-able-to-unsee-58816645.

Kini, Aditi Natasha. 2021. Royal Representation. On the Strange Racial Politics of 'Bridgerton'. *Bitchmedia*, January 14. Accessed August 17, 2022. https://www.bitchmedia.org/article/bridgerton-diversity-colorblind-storytelling-colorbaiting.

Komonibo, Ineye, and Kathleen Newman-Bremang. 2020. A Double Hot Take On *Bridgerton*, Race & Romance. *Refinery 29*, December 28. Accessed March 14, 2022. https://www.refinery29.com/en-us/2020/12/10240235/bridgerton-review-blackness-representation.

Krentz, Jayne Ann. 1992. Introduction. In *Dangerous Men and Adventurous Women: Romance Writers on the Appeal of the Romance*, ed. Jayne Ann Krentz, 1–9. Philadelphia: University of Pennsylvania Press.

Lenker, Maureen Lee. 2020. How *Bridgerton* is Poised to Revolutionize Romance on Television. *Entertainment Weekly*, November 13. Accessed April 26, 2022. https://ew.com/tv/bridgerton-poised-revolutionize-romance-television/.

Mignolo, Walter D. 2018. What Does It Mean to Decolonize? In *On Decoloniality: Concepts, Analytics, Praxis*, ed. Walter D. Mignolo and Catherine E. Walsh, 105–134. Durham: Duke UP.

Offen, Karen. 2007. How (and Why) the Analogy of Marriage with Slavery Provided the Springboard for Women's Rights Demands in France, 1640–1848. In *Women's Rights and Transatlantic Antislavery in the Era of Emancipation*, ed. Kathryn Kish Sklar and James Brewer Stewart, 57–81. New Haven: Yale University Press.

Patton, Venetria K. 2000. *Women in Chains: The Legacy of Slavery in Black Women's Fiction (SUNY series in Afro-American studies)*. New York: State University of New York Press.

Pittman, L. Monique. 2017. Colour-Conscious Casting and Multicultural Britain in the BBC *Henry V* (2012): Historicizing Adaptation in an Age of Digital Placelessness. *Adaptation* 10 (2): 176–191.

Posti, Piia K. 2014. Concurrences in Contemporary Travel Writing: Postcolonial Critique and Colonial Sentiments in Sven Lindqvist's *Exterminate All the Brutes* and *Terra Nullius*. *Culture Unbound* 6 (7): 1319–1345.

Prescott, Amanda-Rae. 2021. Race and Racism in Austen Spaces: Notes on a Scandal: Sanditon Fandom's Ongoing Racism and the Danger of Ignoring Austen Discourse on Social Media. *ABO: Interactive Journal for Women in the Arts 1640-1830* 11 (2): 1–25. Article 10.

Quarles, Benjamin. 1996. Lord Dunmore's Ethiopian Regiment. In *The Negro in the American Revolution*. (1961), 19–32. Reprinted with a new introduction by Gary B. Nash and foreword by Thad W. Tate. Chapel Hill: The University of North Carolina Press.

Roach, Catherine M. 2016. *Happily Ever After: The Romance Story in Popular Culture*. Bloomington: Indiana University Press.

Sanders, Julie. 2006. *Adaptation and Appropriation*. London and New York: Routledge.

Shanks, John. 2022. Presentist Anachronism and Ironic Humour in Period Screen Drama. *Research in Film and History. New Approaches* 1–35. https://doi.org/10.25969/mediarep/18962.

Stam, Robert. 2017. Revisionist Adaptation: Transtextuality, Cross-Cultural Dialogism, and Performative Infidelities. In *The Oxford Handbook of Adaptation Studies*, ed. Thomas Leitch, 239–250. New York: Oxford University Press.

Thomas, Paul L., ed. 2013. *Science Fiction and Speculative Fiction. Challenging Genres*. Rotterdam: Sense Publishers.

Thompson, Ayanna. 2011. *Passing Strange: Shakespeare, Race, and Contemporary America*. New York: Oxford University Press.

Tutan, Defne Ersin. 2017. Adaptation and History. In *The Oxford Handbook of Adaptation Studies*, ed. Thomas Leitch, 576–586. New York: Oxford University Press.

Van Dusen, Chris. 2021. "Bridgerton" Showrunner on Creating A Color-Conscious Series (Guest Column). *The Hollywood Reporter*, August 23. Accessed August 24, 2022. https://www.hollywoodreporter.com/tv/tv-news/bridgerton-showrunner-creating-color-conscious-series-guest-column-1234998873/.

Wilde, Lawrence. 1999. The Radical Appeal of Hermann Hesse's Alternative Community. *Utopian Studies* 10 (1): 86–97.

Wilkins, Kim, Beth Driscoll, and Lisa Fletcher. 2022. *Genre Worlds: Popular Fiction and Twenty-First-Century Book Culture*. Amhearst and Boston: University of Massachusetts Press.

Wood, Marcus. 2002. *Slavery, Empathy and Pornography*. Oxford: Oxford University Press.

Young, Erin S. 2020. Race, Ethnicity, and Whiteness. In *The Routledge Research Companion to Popular Romance Fiction*, ed. Jayashree Kamblé, Eric Murphy Selinger, and Hsu-Ming Teo, 511–528. London: Routledge.

CHAPTER 8

Ted Chiang's Counterphysical Stories and History of Science Pedagogy

John L. Hennessey

Introduction

Could things have been different? We tend to think so, at least when it comes to human decisions and actions. The answer is much less clear when it comes to nature or the universe. Advances in mathematically-based particle physics have given rise to a so-called Fine-Tuned Universe Argument, in which life on Earth could only exist under extremely specific conditions. It is far less likely, or perhaps impossible, that analogous life could have developed if Earth were outside of the so-called Goldilocks zone in its orbit around the Sun, if the Solar System had been located closer to the galactic center, with its higher level of deadly cosmic rays, or if Jupiter had not existed in its relative position to divert most extinction-level asteroids.[1] More broadly, tweak the value of a universal constant, or the initial conditions of the Big Bang, and galaxies, stars, and the relative abundance of certain life-sustaining elements might no longer be possible. Such lines of thought have led to bitterly contested arguments about intelligent design or the existence of our universe as but one instance in a multiverse,

J. L. Hennessey (✉)
Lund University, Lund, Sweden
e-mail: john.hennessey@kultur.lu.se

J. L. Hennessey (ed.), *History and Speculative Fiction*,
https://doi.org/10.1007/978-3-031-42235-5_8

but regardless of one's position on such issues, most physicists seem to agree that life as we know it was made possible by very specific conditions that could hardly have been much different (Landsman 2016).

These disparate ways of thinking about contingency are reflected in literature. Counterfactual histories constitute a prominent subgenre of speculative fiction, imagining how the shape of history would have turned out if, say, Nazi Germany had won World War II, and often dramatizing what this would mean for individuals "on the ground" (see, for example, Gallagher 2018; Hellekson 2001). Nevertheless, in spite of diverging from history *wie es eigentlich gewesen*, such works tend to otherwise be highly realistic. In several of his stories, award-winning speculative fiction author Ted Chiang goes much further, setting his narratives not only in an alternative timeline, but in an alternate reality in which the very laws of nature differ in significant ways. Echoing the established concept of the "counterfactual," this chapter coins the term "counterphysical" to describe this type of literature. Counterphysical literature plays with the very laws of physics or the nature of the universe, but still within a rules-based, science-inspired paradigm that distinguishes it from systems of magic in fantasy literature. The existence of counterphysical literature has been noted in several scholarly works on alternative history fiction, but to my knowledge has received no extensive attention as a subcategory or subgenre of its own (Hellekson 2001, 50, Baxter 2019, 2–3).

This chapter will examine two such stories by Ted Chiang, "Seventy-Two Letters" (2002 [2000]) and "Omphalos" (2019), and argue that they can serve as a useful pedagogical tool for teaching the history of science. In the pedagogical framework "Decoding the Disciplines" developed by David Pace, instructors should identify and actively remedy "obstacles to learning in the discipline and the kinds of mental operations that students must master to overcome such obstacles" (Pace 2017, 4). In the history of science, a crucial but difficult "mental operation" is to suspend one's present-day understanding of science in order to comprehend the worldview and mindset of historical scientists operating in a very different epistemological milieu. I argue that counterphysical fiction like Chiang's can usefully be employed to help train students to overcome such obstacles to thinking about alternative science epistemologies.

Counterphysical Speculation and Its Implications

What I term *counterphysical* literature is far less common and diverges in important ways from the more widespread counterfactual fiction.[2] It involves a setting in which the laws of physics or of nature (whether gravity, electromagnetism, biological processes, and so on) are significantly different from those of our known universe. Beyond this positive definition, it may be most helpful to provide a series of negative definitions, that is, types of fiction that do *not* count as counterphysical.

Most basically, counterphysical literature must go beyond a mere alternate timeline. How would history be different if (most famously) Hitler had died before coming to power, the Confederacy had won the American Civil War, the Soviet Union did not collapse, Chinese had discovered the Americas before Europeans, and so on? Such questions have given rise to a great deal of provocative, creative, and useful speculative fiction, but this nearly always takes place in the same physical universe as our own. Thus, Ward Moore's *Bring the Jubilee* (1953), Philip K. Dick's *The Man in the High Castle* (1962) (despite some speculation on connections between alternate timelines) and Robert Harris' *Fatherland* (1992) count as counterfactual, but not counterphysical, fiction.

Less obviously, I do not count as counterphysical speculative fiction in which a new invention or scientific discovery opens up new realms of possibility or a new understanding of the physical universe. Narratives involving the discovery of time travel, artificial intelligence, faster-than-light travel, new forms of genetic engineering, or whatever, while more or less plausible, nevertheless generally have their implicit starting point in our familiar physical universe. The same is true of technology introduced by aliens, even if it leads to new understandings of reality. Even if reality is revealed to be a simulation, or the laws of physics do not apply on a higher plane that humans have not yet come to understand, or the like, the starting point is still a familiar physical universe. In contrast, in a counterphysical reality, the different physical and natural laws are a priori; they have always existed and (if the author is skillful) are largely taken for granted by the characters that populate it. Like counterfactual literature, the characters in counterphysical literature generally do not understand or notice that something about their universe is "wrong" or "unnatural," allowing the reader to speculate on how life might otherwise be or have been.

A related riff on the previous examples that can be excluded from the definition of the counterphysical is the combination of counterfactual fiction with new inventions or scientific discoveries. William Gibson and Bruce Sterling's *The Difference Engine* (1990) is a well-known example, exploring the question of how history might have been different if computers had been invented in nineteenth-century Britain. As we shall see, this novel bears many similarities to Chiang's "Seventy-Two Letters." But again, rather than taking place in a different physical universe, the Babbage Engines of this alternative timeline are simply invented earlier, even though this of course has significant ramifications of all kinds. A similar example, *Äkta människor* [Real Humans], envisions an alternate contemporary Sweden in which people can purchase highly intelligent "hubot" androids for a variety of uses, sparking a populist political backlash (Hamrell and Akin, 2012–2014). But although the technology does not yet exist, it is certainly well within the realm of imagination and once again originates in the same physical universe as us. Such stories can generate similar reflections on society, life, the universe, and everything, and indeed other examples of Chiang's fiction fall into this category, but counterphysical fiction operates slightly differently.

Lastly, and less unambiguously, counterphysical fiction is characterized by a scientific idiom. This can be difficult to precisely define, just as it can sometimes be difficult to distinguish between science fiction and fantasy, but a general rule of thumb is that counterphysical fiction does not involve magic. Instead, it is interested in questions of alternative physical or natural laws that can be investigated through empirical, scientific methods. Thus, historical fiction with fantastic elements, like Susanna Clarke's *Jonathan Strange & Mr. Norrell* (2004), is not counterphysical (although the depiction of an early-nineteenth-century scientific society for the study of magic at the beginning of the novel comes close). Tales of Jane Austen with zombies or the Middle Ages with dragons can generally be ruled out as they are typically not interested in scientific questions. The most difficult-to-place example with which I am familiar is Brandon Sanderson's *Mistborn* series (first installment 2006). The novels' idiom is unquestionably high fantasy, but the series' world is characterized by a very complex alchemical system in which certain individuals can ingest and "burn" different metals in order to gain various powers. This system is rigorously and predictably regulated by physical laws, and it is difficult to say whether it counts as "magic" or as an alternative physical reality. Even though the series is not concerned with what would generally be characterized as scientific inquiry, the social, economic, political and religious implications of such a world are explored in great detail, just as in Chiang's stories, as we shall see.

While the dearth of scholarship on what I have defined as counterphysical literature makes it difficult to produce a list of examples, in her 2001 book on alternate history fiction Karen Hellekson mentions two works that clearly fall into this category (2001, 50). The first, Philip José Farmer's short story "Sail On! Sail On!" (1979 [1952]), is a dialogue between sailors and a friar on Columbus' *Santa Maria*. These individuals, who live in an alternate fifteenth century marked by a version of wireless communication and a firm belief in the roundness of the earth, discuss the possibility of parallel universes with alternate timelines and even physical laws, before falling off the edge of what turns out to be a flat Earth. Richard Garfinkle's 1996 novel *Celestial Matters* is perhaps the most important contribution to the subgenre, creatively combining both counterfactual and counterphysical speculation. Counterfactually, the story is set in a world that has for centuries been divided between two warring superpowers: the Classical Greek Delian League and China. Christianity has apparently never emerged (although pacifist Buddhists are a thorn in the side of both empires), and the balance of power between different regions of the world is considerably different, with Northern Europeans serving as slaves while many Africans, Indians, and Native Americans are fully assimilated into Greek society. Counterphysically, the universe in which the characters live and operate is consistent with the understandings of the Ancient Greeks from our universe (as well as traditional Chinese medicine and philosophy, as becomes evident later in the book). The celestial bodies, composed of special matter, are set in crystalline spheres, animals arise through spontaneous generation, matter is composed of the four classical elements and medicine effectively makes use of the human body's humors. In his blurb on the book cover, Harry Turtledove describes the book as "hard science fiction," and the novel has been widely praised for its rigorous application of Ancient Greek understandings of the physical world, which are continually important for the novel's plot. Like "Omphalos," examined below, *Celestial Matters* is a useful resource for understanding the full implications of an alternate, discredited set of physical laws, indirectly revealing how we know they are false.

Counterphysical literature, according to this stringent definition, is an unusual subgenre of speculative fiction, but one with a perhaps unique potential to explore deep existential questions about human society and the physical universe. The following section will present Ted Chiang's two stories "Seventy-Two Letters" and "Omphalos" and their counterphysical elements, before subsequent sections develop how these could be used to help students to better grasp the history of science.

Playing with the Laws of Physics: "Seventy-Two Letters" and "Omphalos"

From the beginning, it becomes clear that "Seventy-Two Letters" is set in a version of Victorian England, complete with the industrialization of textile manufacture and its resultant class conflicts, the rapid advance of scientific discovery under the auspices of the Royal Society, and even the Crimean War. Nevertheless, it also becomes apparent almost immediately that the world of this Victorian England is subject to vastly different physical and natural laws than our reality. There are three major differences in Chiang's alternative England: human reproduction occurs very differently, the origin of species and their evolution is decidedly non-Darwinian, and most notably, seventy-two-letter written "names" contain the power to animate dolls formed of metal or clay and affect other aspects of the physical world, such as human health or heat transfer. On top of this, in only around sixty pages, Chiang manages to combine a gripping narrative with deep reflections on class, industrialization, the relationship between science and religion, medical ethics, eugenics, and reproductive freedom. It is an astoundingly unique literary achievement.

Treating one counterphysical element at a time, we are introduced to this Victorian England's peculiar mode of human reproduction when the protagonist, while still a schoolboy, is confronted with a backyard experiment undertaken by one of his friends, in which he has incubated and grown the embryos in his sperm. Without any indication of the strangeness of this experiment, we learn that in this reality, the entire line of a species is simultaneously created and stored in the males' sperm; every man literally contains all of his descendants within himself. Nevertheless, although containing the basic human form and substance necessary for life, the sperm cannot "quicken" by itself nor take on any unique characteristics (physical or mental), until animated by the "vital force" provided by the mother in her ovum. The implications of this counterphysical reproductive order become clear when the protagonist is informed of secret experiments conducted by the Royal Society and its French counterpart that have revealed that the line of embryonic generations contained in all human males will soon come to an end; soon, all men will become infertile and the human race will go extinct. The protagonist and his colleagues propose a bold solution to this existential threat to humanity, but one which raises stark ethical questions about reproductive ethics, socioeconomic equality, and human freedom.

Although not a primary focus of the story, the discussion of the future of the human race and artificial means of reproduction naturally leads to a consideration of the origin of species and their extinction. The fossil record in this reality indicates that species do not change over time, but rather appear suddenly and eventually go extinct. We learn that although a familiar "Catastrophist" explanation of species extinction has been proposed, this appears less probable in light of the discovery of finite generational lines of fetuses. Instead, major catastrophes are postulated as the possible origin of life (the trustworthy elder scientist who explains this also takes the spontaneous generation of simple organisms as a matter of fact) (166). Thus, the origin and development of species and interpretation of the fossil record is very much on the agenda of these alternative Victorian scientists, even though the empirical facts they produce are strikingly different than those proffered by Darwin and others in our own reality.

The powerful "names" that animate dolls and other "engines" are unquestionably the most dramatically counterphysical aspect of the story. The seventy-two letters of the story's title are typically written in Hebrew on slips of paper that are then inserted into prepared "automata," usually clay or metal dolls with a more or less human or animal form. One inside, these "names" somehow touch the essence of the form that is being animated, and careful combinations of distilled terms and synonyms prepared by "nomenclators" can give the automata increasing dexterity and ability.

This clearly differs sharply from the physical laws of our universe, but Chiang once again explores the socioeconomic, cultural, and even religious consequences of a universe in which names are imbued with power. Automata are put to work as children's toys, carriage-pullers, engines powering spinning factories (rendering steam technology superfluous), and even animated sex dolls. The best names are protected by patents, and the development of increasingly able automata leads to worker unrest, as skilled tradesmen fear for their jobs. Most interestingly, the power of names has long been seamlessly integrated into Christianity in this alternate Victorian England. A Biblical reference to an automata sex doll made by Jacob's sons makes it clear that the universe of the story has always been different from the reader's own (177). Names are still written in Hebrew letters, and their power is traditionally explained as their reflecting God's name or the names He granted to His creation (148). A younger generation of nomenclators and scientists has started to question this reasoning, however, seeking instead a secular explanation for the power of names (149–150). Even in the absence of Darwin and the steam engine, this Victorian England increasingly moves towards secularism and industrialization. Here, as elsewhere in Chiang's speculative fiction, the reader is left to wonder whether apparently key historical developments or natural laws were really necessary to the broad strokes of modern history or whether these would have happened anyway.

The same question arises in Chiang's later story "Omphalos," which takes place in a world at once startingly similar to and shockingly different from our own. The premise of this counterphysical world is less original but no less compelling: it is a creationist Earth, in which the physical fingerprints of God's creation are readily apparent to empirical scientists. Archaeologists have uncovered a great deal of "primordial" fossils and other remains of God's original creation that are collectively referred to as "relics," whether trees whose rings stop at a certain point (the core of the trunk being homogenous at the size in which the adult tree was brought into being), mummified humans without navels or adult animal bones without lines where they fused together in infancy. Using various scientific methods, scientists have been able to determine that the Earth's age is exactly 8912 years (240). Nor is this all; there are exactly 5872 stars in the sky, all of which are "identical in size and composition" (254). It is taken for granted that the Earth is the center of the universe, hence the story's title, although this is eventually challenged in an unexpected twist that remains true to the story's reality. The universe of "Omphalos" is thus very similar to familiar ancient and medieval understandings of our universe; only in the story, these can be corroborated with modern science.

Naturally, a great deal of this story deals with the relationship between science and religion, as well as their implications for the purpose of humanity. But just as in "Seventy-Two Letters," society in the world of "Omphalos" is not as different as we might expect, given the vast differences in physical and natural laws. Scientists, like the archaeologist protagonist, are typically deeply religious individuals seeking to reveal God's purpose for the world through their empirical work, and much of the story is narrated in the form of this character's prayers. Of course, even in our reality, many scientists have such motives, but "Omphalos" has few indications of any religious-secular divide in society or the academy. Nevertheless, even in a world with such unambiguous evidence of intelligent design, the pious narrator and others are constantly concerned about people straying from God's path and feel a strong need to invigorate their faith, whether through scientific lectures about God's creation or the experience of coming into contact with primordial relics. Nor is the interpretation of God's will unambiguous and unchanging. Indeed, the purpose of science, for the protagonist, is to better understand God's intentions with creation. As a female scientist, she rejects the church's traditional teachings in which "every woman… continue[s] to live in Eve's shadow," pointing out that science had disproven the Biblical account of human creation (apparently the same as ours) by showing that humans were created simultaneously around the world (264). The Church in this reality, like many Christian churches in ours, had accepted the scientific evidence and decided to reinterpret the story of Adam and Eve as an allegory (264). In short, the relationship between science and religion seems to be less fraught in this creationist universe, but still not completely straightforward.

"Seventy-Two Letters" and "Omphalos" are both clear examples of what I have described above as "counterphysical" literature. Their realities differ from ours in fundamental, obvious ways, whether through the power of names to animate clay dolls, the presence of all future generations in each male individual or unambiguous evidence that the world and its inhabitants were created fully formed less than 10,000 years ago. Nevertheless, in Chiang's telling, these factors do not cause human history or society to diverge in terribly significant ways from those familiar to the reader. This is not for a lack of exploration of the social, economic, and cultural consequences of, for example, name-animated "engines" or an empirically-backed theology. Like all speculative fiction, Chiang's stories invite the reader to scrutinize the believability of the alternative reality that they offer, which I will discuss more below. First, however, it is necessary to briefly return to the "real world" and discuss the field of history of science education to which I argue counterphysical literature can be fruitfully applied.

Decoding the History of Science

The history of science as a field of study has expanded dramatically since the mid-twentieth century, but even as it has become an increasingly popular subject for students, there has not been an equivalent expansion of research into history of science pedagogy. What research exists on methods of history of science education has been dominated by discussions of the need to integrate the history of science into the curriculum of the natural sciences and how best to do so (see, for example, Gooday et al. 2008; Kolstø 2008; Duschl 2006).[3] Here, I will argue that the Decoding the Disciplines Paradigm developed by historian David Pace (2017) can fruitfully be applied to the teaching of the history of science, before connecting this to counterphysical literature in the following section.

Pace's starting point is the insight that "Knowing how to do something is a different thing than knowing how to teach that thing" (2012). Like other higher education researchers, Pace argues that being an expert in an academic field is insufficient for being a good teacher—instead, it is necessary to critically reflect on students' learning process and needs and apply the findings of pedagogical research. Pace's Decoding the Disciplines Paradigm distinguishes itself from other pedagogical approaches through its adherence to three basic principles: (1) many questions of learning are discipline-specific, and it is at the level of the discipline that the most detailed pedagogical reflection should be conducted; (2) it is more productive to "concentrate on what students have to *do*, not what they have to *know*"; and (3) how to perform the most basic tasks necessary to a discipline is not self-evident, but may have become invisible and taken-for-granted by specialists in the field, necessitating very deliberate reflection on these processes (2017, 4–5, original emphasis). Decoding the Disciplines is therefore oriented toward making explicit and modeling the detailed "mental operations" and other steps necessary to a specific discipline.

Pace next details "seven steps of decoding" that the instructor should use to operationalize this educational philosophy (2017, 6):

1. Identify a bottleneck (as Pace elsewhere puts it, "Where in my courses do many students consistently fail to master crucial ideas or actions?" (2012, 50))
2. Define the mental operations needed to get past the bottleneck.
3. Model these tasks explicitly.
4. Give students practice and feedback.
5. Motivate the students and deal with potential emotional blocks.
6. Assess how well students are mastering the mental operations.
7. Share what you have learned about your students' learning.

Here again, Decoding the Disciplines shows itself to be a very hands-on, problem-solving approach to education, seeking out the detailed steps that students need to learn and hammering these home with an almost engineering mentality.

Considering all of these steps in detail in the context of teaching the history of science could easily take up several articles of its own, so here I will focus on one common "bottleneck" in this field: understanding the often unfamiliar social and epistemological context and worldview in which past scientists operated. This is one of the signature questions of the history of science, and many of its best-known classics, like Thomas Kuhn's *The Structure of Scientific Revolutions* (1962) and Steven Shapin and Simon Schaffer's *Leviathan and the Air-Pump* (2011 [1985]), center on such issues. Both of these address widespread misconceptions about the field as telling a heroic, teleological story of universal scientific truths being steadily uncovered, a journey from ignorance to enlightenment and the "right" way of understanding the universe. As Kuhn notes at the beginning of his book, the tendency to view the history of science as a movement from superstition *to* "science" was increasingly questioned even at the time of writing in the 1960s:

> historians confront growing difficulties in distinguishing the "scientific" component of past observation and belief from what their predecessors had readily labeled "error" and "superstition." The more carefully they study, say, Aristotelian dynamics, phlogistic chemistry, or caloric thermodynamics [all discredited systems], the more certain they feel that those once current views of nature were, as a whole, neither less scientific nor more the product of human idiosyncrasy than those current today. (1996, 2)

Kuhn's discussion of phlogiston theory, an explanation for combustion and other processes in terms of an element ("phlogiston") contained inside combustible materials that was released when they were burned, and its subsequent supersession by a new oxygen paradigm, requires a detailed understanding of how the proponents of phlogiston viewed the world. It is far from sufficient to simply label them as "ignorant" or "unenlightened"; they must be assumed to be intelligent individuals and their reasons for believing in phlogiston must be properly considered. Similarly, Kuhn notes that our current scientific theories are not always completely satisfying, pointing out the resistance to ideas of fundamental forces by many historical scientists because they lacked a satisfactory explanation, rather seeming to attribute an "occult quality" to matter (1996, 105). Was it so irrational for scientists to express skepticism of understandings of gravity as an innate, invisible force whose mechanism of operation was unknown?

Shapin and Schaffer's book problematizes the teleological, heroic narrative of scientific progress in similar ways. By scrutinizing Robert Boyle's famous air pump experiments and their contemporaneous reception, the authors demonstrate that Boyle's experimentally based insights were far from obvious truths that immediately gained widespread acceptance. Rather, making these into generally accepted "matters of fact" required both a great deal of work (including practical work on the air pump apparatus, to ensure the credibility and reliability of Boyle's experiments) and a specific kind of expert community that could evaluate and vouch for Boyle's results and interpretations (2011, 225). For a wide variety of reasons, alternative explanations to Boyle's for the behavior of air have been discredited, but it is difficult to understand this without suspending one's present-day scientific knowledge and seriously considering the positions of Boyle's opponents. Shapin and Shaffer list a large number of often distinguished studies of Boyle and his opponent Thomas Hobbes that either ignore Hobbes' scientific writings (concentrating on his political ones) or else dismiss these as based on "misunderstanding" without a close reading (2011, 12).

Perhaps even more insightfully, Shapin and Schaffer point out that most "historians start with the assumption that they (and modern scientists) share a culture with Robert Boyle, and treat their subject accordingly," a tempting fallacy (2011, 5). In fact, even though his theories are currently recognized as generally correct, it is a mistake to view Boyle as thinking and operating according to our own modern worldview, a fallacious approach that could easily lead to misunderstanding or missing important aspects of Boyle's thought. Shapin and Schaffer express the crux of this key mental operation necessary for historians of science when they assert that "We wish to adopt a calculated and an informed suspension of our taken-for-granted perceptions of experimental practice and its products" (2011, 6). They suggest that researchers proceed by "playing the stranger": "one great advantage the stranger has over the member in explaining the beliefs and practices of a specific culture: the stranger is in a position to *know* that there are alternatives to those beliefs and practices" (2011, 6).

Needless to say, "playing the stranger," or accomplishing a "suspension of our taken-for-granted perceptions" is challenging even for professional researchers, even as it is arguably crucial for students of the history of science. For this mental operation is necessary to avoid the common but fallacious view of science as inevitably moving toward present-day understandings and to be able to comprehend now-discredited theories without merely dismissing them as erroneous. Identifying this common "bottleneck" in students' understanding in the history of science is one of the first steps in the Decoding the Disciplines paradigm, but what strategies can teachers use to help students move beyond it? The following section will argue that by plunging students into an alternate reality, counterphysical fiction could help students to practice "playing the stranger," training them for a more nuanced understanding of the ("real") history of science.

Counterphysical Fiction as a Pedagogical Tool

Speculative fiction is clearly useful for more general reflecting on large philosophical questions like the nature of the universe and human society, but how can it be connected to such a practical, step-by-step pedagogy as Decoding the Disciplines? In the case of counterphysical fiction, at least, I argue that its exploration of the consequences of an alternate universe with different natural and/or physical laws is singularly useful for illustrating and practicing the "mental operations" necessary to the history of science. For the historical scientists at the core of its curriculum often had radically different understandings of reality, which in some ways resembled a (from our vantage point) counterphysical universe.

In an important book on counterfactual history and fiction, Catherine Gallagher writes that one significant role played by such speculation is that "it helps satisfy our desire to quicken and vivify historical entities, to make them seem not only solid and substantial but also suspenseful and unsettled" (2018, 11). This is a useful description of the very mental operation required by the history of science discussed here: making the history of scientific "discovery" more "suspenseful and unsettled" by showing how it was affected by social factors (and not only empirical "truth") and could have been (and indeed often was) theorized differently. By launching us into such a "suspenseful and unsettled" (and indeed, often unsettling) alternate reality, counterphysical fiction is a useful tool for destabilizing our taken-for-granted understandings of the physical world.

In so doing, counterphysical fiction naturally invites epistemological reflection: How do we know what we know? How can we be so sure that the universe operates the way we think it does? What aspects of the natural and physical world remain unknown and mysterious? These are of course fundamental questions for the philosophy of science and for understanding the history of the scientific method and the curiosity that drove scientific work both in the past and today. And indeed, both of Chiang's stories devote considerable, explicit attention to epistemological questions. While the power of names in "Seventy-Two Letters" is mostly taken for granted, the pending infertility of male humans is a new scientific discovery, and thus elicits a long discussion of how the scientists at the Royal Society and across the Channel can be so sure of their conclusion. Similarly, "Omphalos" is largely about scientific epistemology, as it centers around the production of evidence of God's creation of the universe. Nevertheless, the meaning of even such seemingly unambiguous evidence as navel-less mummies is to a great extent open to interpretation, as the characters discover at the story's end. When compared to actual examples from the history of science, to what extent are the different scientific worldviews a product of different empirical facts, and to what extent do these depend on the human context of their production?

How is reflecting on these questions using counterphysical fiction different than merely employing historical examples of alternative, now discredited, scientific theories and ways of understanding the world? The very idea of a fictional world with different natural or physical laws can provide an alternative path into reflections that are normally shut down by the strong bias that discredited theories are merely wrong or "unscientific." By positing a world in which the reality of physical and natural laws actually *is* different from ours, a series of "mental operations" is set in motion that can then be transposed to real-world historical cases. Counterphysical fiction helps us to suspend our biases about the nature of the universe (not unlike the "suspension of disbelief" important so speculative fiction overall) in a similar way to what is necessary to understand historical scientists in their own contexts (as best as we can, at any rate). Reading and reflecting on stories like those of Ted Chiang in history of science classes can therefore be a kind of practice for "the real thing" and provide interesting examples that can later be productively compared with historical cases. For example, how did creationist theories and religious reactions to Darwinism in different periods differ from what Chiang imagines would have been the case in a world in which all the scientific evidence in fact pointed to divine creation?

An important difference between counterphysical fiction and superseded historical scientific theories is that fictional worlds are typically not as messy as the real one. In both his stories, Chiang is able to create alternative, but coherent realities that his characters can investigate rationally. In contrast, as Kuhn notes, although "normal science" may seem to be "a single monolithic and unified enterprise that must stand or fall with any one of its paradigms as well as with all of them together," actually "science is obviously seldom or never like that. Often, viewing all fields together, it seems instead a rather ramshackle structure with little coherence among its various parts" (1996, 49). No short story could capture the complexity of the "ramshackle structure" that comprises science both historically or today, and such fiction may risk reinforcing oversimplified understandings of science history. If we are to make a virtue of necessity, however, counterphysical fiction could also be seen as providing an easy point of entry to a difficult subject. Just as it may assist students in learning to temporarily suspend their biases about past theories, fiction like Chiang's stories could also provide an easier practice run before studying the convoluted incoherence of historical scientific debate.

I envision history of science classes beginning with a reading and detailed discussion of counterphysical fiction as a useful point of entry to demanding investigations of historical scientific theories and worldviews. Students would be trained to ask the kinds of questions necessary to the discipline of the history of science about epistemology, scientific methods, and the relationship between social context and science without the same risk of these being shut down by a teleological narrative of scientific progress in which alternate theories are to be discarded as simply false. Understanding the context in which these theories arose, and gauging their explanatory power, while remaining cognizant that they have been disproven by future scientific findings, would help the student to understand the processes by which more successful scientific theories also came into being and gained widespread acceptance.

Conclusions

This chapter has covered much ground: it began by defining a new subgenre of speculative fiction, counterphysical fiction, described two examples of this by Ted Chiang, and then argued for the usefulness of such fiction for history of science pedagogy within the Decoding the Disciplines educational framework. In short, I contend that the type of epistemological reflections that counterphysical fiction, set in an alternate universe with different physical and\or natural laws, invites the reader to partake in, provide useful training for critical thinking within the history of science. Counterphysical fiction forces the reader to be a stranger in a strange world, providing practice for thinking of historical science in similar ways rather than in ways overdetermined by our familiarity with subsequent developments of modern science.

This example of the usefulness of a subgenre of speculative fiction for a subdiscipline of history is but one case of how speculative fiction and history can benefit from close encounters. And indeed, this benefit is often mutual. Although not discussed explicitly above, it is of course the case that Chiang's counterphysical fiction has also been heavily inspired by the history of science, as well as social history more broadly. Chiang's familiarity with the historical structures of scientific knowledge-making greatly enriches his stories' settings, while allowing him to raise provocative questions about society in the limited space provided by the short story form.

Notes

1. Astrophysicist Neil F. Comins explores a variety of such hypothetical scenarios in his books *What If the Moon Didn't Exist?* (1993) and *What If the Earth Had Two Moons?* (2010), but all of his "what if" scenarios follow the known physical laws of the universe.
2. A more detailed breakdown of counterfactual fiction into different subtypes can be found in Gallagher 2018, 2–3.
3. A notable exception is Hendriksen 2020.

References

Baxter, Stephen. 2019. Foreword. In *Sideways in Time: Critical Essays on Alternate History Fiction*, ed. Glyn Morgan and C. Palmer-Patel, 1–10. Liverpool: Liverpool University Press.

Chiang, Ted. 2002. *Stories of Your Life and Others*. New York: Tor.

———. 2019. *Exhalation*. New York: Knopf.

Clarke, Susanna. 2004. *Jonathan Strange and Mr Norrell*. London: Bloomsbury.

Comins, Neil F. 1993. *What if the Moon didn't Exist?* New York: HarperCollins.

———. 2010. *What if the Earth had Two Moons?* New York: St. Martin's Press.

Dick, Philip K. 1962. *The Man in the High Castle*. New York: Putnam.
Duschl, Richard A. 2006. Relating History of Science to Learning and Teaching Science: Using and Abusing. In *Scientific Inquiry and Nature of Science*, ed. Lawrence B. Flick and Norman G. Lederman, 319–330. Cham: Springer.
Farmer, Philip José. 1979 [1952]. Sail On! Sail On! In: *The Road to Science Fiction*, vol. 3, ed. James Gunn, 195–204. New York: Mentor.
Gallagher, Catherine. 2018. *Telling It Like It Wasn't: The Counterfactual Imagination in History and Fiction*. Chicago: University of Chicago Press.
Garfinkle, Richard. 1996. *Celestial Matters*. New York: Tor.
Gibson, William, and Bruce Sterling. 1990. *The Difference Engine*. London: Gollancz.
Gooday, Graeme, John M. Lynch, Kenneth G. Wilson, and Constance K. Barsky. 2008. Does Science Education Need the History of Science? *Isis* 99 (2): 322–330. https://doi.org/10.1086/588690.
Hamrell, Harald, and Levan Akin, dir. 2012–2014. *Äkta människor* [Real Humans]. Stockholm: Sveriges Television.
Harris, Robert. 1992. *Fatherland*. London: Arrow Books.
Hellekson, Karen. 2001. *The Alternate History: Refiguring Historical Time*. Kent, OH: Kent State University Press.
Hendriksen, Marieke M.A. ed. 2020. Rethinking Performative Methods in the History of Science (Special Issue). *Berichte zur Wissenschaftsgeschichte* 43. https://onlinelibrary.wiley.com/toc/15222365/2020/43/3.
Kolstø, Stein Dankert. 2008. Science Education for Democratic Citizenship through the Use of the History of Science. *Science & Education* 17: 977–997. https://doi.org/10.1007/s11191-007-9084-8.
Kuhn, Thomas. 1996 [1962]. *The Structure of Scientific Revolutions*. 3rd ed. Chicago: University of Chicago Press.
Landsman, Klaas. 2016. The Fine-Tuning Argument: Exploring the Improbability of Our Existence. In *The Challenge of Chance*, ed. Idem and Ellen van Wolde, 111–129. Cham: Springer.
Moore, Ward. 1953. *Bring the Jubilee*. New York: Farrar, Straus & Young.
Pace, David. 2012. Decoding Historical Evidence. In *Enhancing Student Learning in History*, ed. David Ludvigsson, 49–62. Uppsala: Opuscula Historica Upsalensia.
———. 2017. *The Decoding the Disciplines Paradigm: Seven Steps to Increased Student Learning*. Bloomington: Indiana University Press.
Sanderson, Brandon. 2006. *Mistborn: The Final Empire*. New York: Tor.
Shapin, Steven, and Simon Schaffer. 1985. *Leviathan and the Air-Pump: Hobbes, Boyle and the Experimental Life*. Princeton: Princeton University Press.

CHAPTER 9

The Dark Past of Our Bright Future: Concurrent Histories of *Star Trek: Voyager*

Ella Andrén

"Captain's log, stardate 53896." The *Star Trek: Voyager* episode "The Muse" (season 6, episode 22) starts with the same words that so many episodes of the *Star Trek* franchise do, the beginning of a dated log entry. However, this is not the usual voiceover by a familiar character of the series. Here, the words are recited in unison by what looks very much like a choir from Ancient Greek theatre. The actors are dressed in robes, carrying masks on sticks, and perform, lit by torchlight, in a circular space quite reminiscent of the ancient amphitheater. A wealthy patron holds a central seat in the audience, enjoying the show.

But this is not Earth in ancient times. This is, in fact, the future on an alien planet, somewhere in the Delta Quadrant, the far side of the Milky Way galaxy, where the starship *Voyager* is trying to make its way back to Earth. In "The Muse," *Voyager*'s chief engineer B'Elanna Torres has crashed on this planet, and while she's been down with a fever, the resident Kelis the Poet has found inspiration for a play in her shuttlecraft's logs. As his patron, on whose charity the actors depend, has very much enjoyed the resulting play, "The Away Mission of B'Elanna Torres," Kelis

E. Andrén (✉)
Linnaeus University, Småland, Sweden

J. L. Hennessey (ed.), *History and Speculative Fiction*,
https://doi.org/10.1007/978-3-031-42235-5_9

the Poet is desperate to learn more about what he believes are gods, the "Voyager Eternals," in order to write a subsequent play about them. In fact, his patron demands it in just one week.

This episode is a clear example of how layered *Star Trek: Voyager* is with time. Past, present, and future are intertwined. Through the futuristic visions of Star Trek, the imagined future and past can exist concurrently. Starfleet's present, out imagined future, can encounter another civilization, quite similar to our own past, which is thereby also their past. In the *Voyager* episode "The Muse," the ancient Greek-like actors perform their own version of the well-known science fiction story, struggling with the demand of creating new adventures every week, as well as with trying to understand the alien characters and their actions, in a joyful play with the actual series' real-life audience and their demand for weekly new space adventures.

In this chapter, I mean to show how *Star Trek: Voyager* makes use of history, focusing on this metaphorical way of making the past concurrent with the present and future.

The Future Is Televised

The Star Trek franchise is frequently described as utopian, the bright future of humanity, Earth, and the universe. Various series and movies of the franchise are set mainly in the twenty-third and twenty-fourth centuries, but venture from the twenty-second century (*Star Trek: Enterprise,* 2001–2005) and into the thirty-second (*Star Trek: Discovery*, 2017–). The timeline of the original Star Trek (1966–1969) has been expanded with both prequels and sequels, spanning a multimedia franchise of TV series, movies, animation, comics, books, games, and toys.

The time and society of the Starfleet explorers and the United Federation of Planets, which enlists them, is said to have overcome issues of inequality, racism, sexism, hunger, war, capitalism, greed, and environmental problems. (How all these pressing issues have been resolved is not always made clear, but throughout the franchise, bits and pieces of this development are given.) This is at the same time a radical fantasy of the future and an artistic and commercial product of our own time. In their travels across the galaxy, the Starfleet explorers also encounter different civilizations, often depicted or even described as reminiscent of various time periods of Earth's history.

In *Star Trek: Voyager*, history is present in occasional time-travel episodes where the viewer is actually taken back to our own time, our past, or that past of the series that would simultaneously be our own future. The past is also well represented in what in *Voyager* are referred to as holonovels: interactive, holographic projections used for entertainment, training, or education. The third use of the past might, as exemplified above, be described as metaphorical. This is where the present of an encountered civilization bears a striking resemblance to something in the protagonists' past, which may be our own past or present. In this way, past, present, and future exist concurrently, and the past is made both vivid, close and alien.

Popular Culture as Historical Culture

The Popular culture theorist John Fiske claims that a writer of popular culture "does not put meaning into the text, but rather assembles a multitude of voices inside it," where

> different readers can "listen" more or less attentively to different voices. The reader makes his or her text out of this "weaving of voices" by a process that is fundamentally similar to that of the writer when s/he created the work out of the voices available in the culture. (Fiske 2001, 96)

In other words, a popular work of fiction is never just one thing, never unambiguous. In order to be popular, it needs to speak with many voices, enabling different audience members to hold their own interpretations of the text.

Historical culture, too, is an incongruent, changing thing. It is often described as both the physical products and the collective ideas of, most notably, a national notion of history. However, even in authoritarian societies, where the official version of the past is made very clear, there are bound to be conflicting notions, counter-narratives, oppositional historical sub-cultures. As George Orwell put it, "Who controls the past controls the future. Who controls the present controls the past" (Orwell 1949, 44). The narration of history is bound to be a struggle for power and meaning.

What does that mean for the use of history in popular culture? Perhaps that the elements of history referred to in works of popular culture need to be easily recognizable to many. More than just past, they need to be History. The amount of conflict, of counter-narrative, is thereby restricted.

In this way, popular imagination might be limited by historical culture, the way Fiske refers to the "voices available in the culture" to both writer and reader in the creation and interpretation of a text. With Star Trek, this might explain why the interstellar Starfleet adventures reference almost exclusively American/Western historical events and phenomena; a different take would not make it recognizable as History.

A Future Stuck in the Past?

Previous scholars writing on Star Trek and history have focused mainly on how the franchise, in particular what is now known as *The Original Series* (*Star Trek,* 1966–1969), reflects the issues and values of its production time, the late 1960s. As just one example, Mike O'Connor concludes, in his article "Liberals in Space: The 1960s Politics of Star Trek," that *The Original Series* "brought into relief both the most noble and the most contradictory aspects of late 1960s liberalism" (2012, 199).

In *The Voyages of Star Trek: A Mirror of American Society Through Time*, K. M. Heath and A. S. Carlisle describe how the Star Trek franchise, from *The Original Series* up to the second season of *Star Trek: Discovery*, mirrors the eras in which it is produced. Using quantitative methods, they measure representation of gender, race, and age and conclude that despite Star Trek's ideals of equality, it is not until *Star Trek: Voyager* that women come close to the same amount of screen time—and authority—as men. They connect this to the advances made by women in the late 1990s, mainly professional advances in fields like politics, the military, the space industry, law, and science. Heath and Carlisle also describe the "enemies" of the *Voyager* crew in terms of environmental issues, healthcare issues, and hopes and fears regarding rapid technological development, especially in AI and holograms (Heath and Carlisle 2020).

The same kind of limitations in screen time and considerations regarding the depiction of women in Star Trek are discussed in "'Fashion's Final Frontier': The Correlation of Gender Roles in Star Trek" by Katharina Andres (2013). In *The Original Series*, Andres point out, the women's dresses—"mini-dresses, with an asymmetrical plunging neckline and go-go boots"—separate them as "'less' functional uniforms than the shirt and pants of the men, which implies that women are 'less' functional as well." Even in the Star Trek series up to and including *Enterprise* (2001–2005), she points out, where uniforms are generally unisex, there is always one recurrent female character who "stand[s] out because of their skintight jumpsuit, which they wear opposed to the official uniform." She concludes: "The sole purpose for the attire seems to be Star Trek's continuing effort to please a stereotypical male audience, proving that in the end, gender equality has not been achieved, not even in the utopian future that Star Trek wants to present" (Andres 2013, 648).

On a similar note, Brian L. Ott and Eric Aoki question Star Trek's politics on race and identity in "Popular Imagination and Identity Politics: Reading the Future in *Star Trek: The Next Generation*" (2001). Unlike what I emphasize in this chapter, Ott and Aoki see alien species in *The Next Generation* as Others. They point out how the Enterprise crew is "coded principally as White" and interpret the recurring alien species of Klingons as "an amalgamation of the coarsest Black stereotypes" and Ferengi as "an amalgam of grotesque anti-Semitic stereotypes" (Ott and Aoki 2001). Although this is certainly a possible interpretation, and although such destructive stereotypes should be taken very seriously, I find Ott and Aoki's reading slightly problematic. If any sign of anger from a Black person is interpreted as stereotypically Black, will it ever be possible to depict Black anger? If greedy characters are automatically perceived as Jewish, does that not just reinforce the stereotype (and make it impossible to have a constructive discussion of greed)? Going back to Fiske's "multitude of voices," though, these stereotypes are inevitably present, available in our culture.

A different take on *Star Trek* and history can be found in Victor Grech's article "The Banality of Evil in the Occupation of Star Trek's Bajor" (2020). Grech approaches war crimes committed by Cardassian officials in their occupation of the planet Bajor by comparing several portraits of these Star Trek perpetrators with Hannah Arendt's analysis of the high-ranking Nazis Adolf Eichmann and Josef Mengele. Grech studies two story-lines centering these Cardassians, one in *Deep Space 9* and one in *Voyager*. On the *Deep Space 9* space station, located near Bajor and the Cardassian border, a former forced labor camp commander is captured and interrogated. On *Voyager*, the Doctor, in a medical predicament, turns to a holographic version of a Cardassian researcher for help, but is soon made aware that the exobiologist has made his medical advancements through cruel experiments on Bajoran prisoners of war (Grech 2020). In Grech's article, the Cardassian evils become a metaphor of human evils, not simply by mirroring the epoch in which the series was produced but by choosing important themes from human history. This, as I have already stated, I find to be a constructive mode of interpretation for many of the themes depicted on *Star Trek: Voyager*.

Voyages into the Unknown (Yet Familiar)

In the rest of this chapter, I focus on aspects of *Star Trek: Voyager* (1995–2001). This series starts with Captain Kathryn Janeway being given command of the starship *Voyager*, a new, state-of-the-art Starfleet vessel. She recruits, among others, the pilot Tom Paris, and the crew goes on *Voyager*'s first mission: to capture a Maquis vessel, where Janeway's security officer, a Vulcan named Tuvok, is working undercover with the rebel group. As *Voyager* approaches the Maquis ship, they are both swept away to the Delta Quadrant, a faraway part of the galaxy, by an alien entity called the Caretaker.

As both crews take heavy casualties, and the Maquis ship is damaged beyond repair, they reluctantly decide to join forces on *Voyager*. The Maquis leader Chakotay becomes Janeway's second-in-command, and half-human, half-Klingon Maquis B'Elanna Torres is made Chief Engineer. The crew is also joined by a couple of aliens indigenous to the Delta Quadrant, an Ocampa called Kes and a Talaxian called Neelix. As the ship's doctor is killed, the emergency medical hologram, an AI holographic projection known only as the Doctor, becomes a permanent member of the crew.

In the beginning of the fourth season of the series, the crew is joined by Seven of Nine, a human assimilated as a child into the Borg, a cybernetic collective composed from different assimilated species linked together into a single hive mind. As Seven of Nine is severed from the Borg collective, Captain Janeway takes her in, set on helping the somewhat reluctant Seven to regain her humanity.

The journey from the Delta Quadrant back to Earth is estimated to take 75 years. As the crew embarks on this long journey, they encounter various kinds of civilizations and alien species. A few of them are familiar to the crew (and the regular Star Trek viewer), but since the Delta Quadrant is unknown territory to Starfleet, many encountered species are never before seen or heard of. In this chapter, both some new and some familiar alien acquaintances will have important roles to play. First, however, we shall take a look at the actual concept of historical change and development.

Time Templates

In the end of one of the first *Voyager* episodes, "Time and again" (season 1, episode 3), *Voyager* comes across an "M-Class planet". (In fact, the crew spent the entire episode on the planet, preventing a catastrophe they were actually the cause of, then time warped back to when they need not set foot there, but all that is beside the point for this chapter.) Throughout the Star Trek franchise, an M-Class planet is a planet with conditions similar to Earth. The fact that the planetary type most familiar to humans is not designated "A-Class" indicates that this scale was not created with Earth as its norm. The diegetically earlier (but from a later television production) human space travelers of *Star Trek: Enterprise* instead use "Mishara-Class planet," indicating that the concept originates from a Vulcan word. This would make sense, diegetically, since the Starfleet founding member planet Vulcan traveled between solar systems before humans did.

In the *Voyager* episode "Time and again," however, the concept M-Class planet is not the only Star Trek or Starfleet notion referred to. When the planet shows up on *Voyager*'s sensors, Tactical Officer Tuvok describes it like this: "Sensors do show humanoid life. There is no satellite system, and no indications of spacecraft in the vicinity. It appears to be a pre-warp civilization." At this, Captain Janeway turns to Neelix, a Delta Quadrant native who just joined the *Voyager* crew, and explains: "Which means, as a policy, we don't involve ourselves in their affairs," to which Neelix responds: "Of course. A most enlightened philosophy."

This policy is known as the Prime Directive and is well established in the Star Trek franchise (both in in-world Starfleet regulations and in series plot writing). According to the Prime Directive, Starfleet is not to interfere with alien cultures who lack the technological development of the interplanetary Federation. A "pre-warp civilization," as cited above, is specifically a civilization that lacks the ability to travel at warp speed, the kind of speed that makes the relatively quick interplanetary travel of Star Trek possible. The moral implications of the Prime Directive, and the reasons for occasionally breaking it, can be discussed at length, but what is interesting in the context of this chapter is what this indicates with regard to historical development.

First, the focus on technological development might not come as a great surprise in a science fiction series. However, Star Trek is a franchise mainly, if not solely, based on liberal, humanitarian values, a series that more often than not points to the value of diplomacy, of learning, meeting, investigating, and understanding the Other, even when they choose to live their lives differently than We do. When speaking of their own progress, Starfleet officials emphasize how humans in their time have eradicated things like war, poverty, and injustice. Captain Janeway makes this clear, for example, when the *Voyager* crew encounters a group of humans in the Delta Quadrant who are the descendants of individuals secretly abducted by aliens in 1937. She muses on their "thriving, sophisticated culture" and concludes:

> The remarkable thing about the humans on this planet, is that they evolved very much like the people of Earth. Tens of thousands of light-years apart, both civilizations managed to create a world they could be proud of, one where war and poverty simply don't exist.

She does call it remarkable, not in any way inevitable. Yet, the impression of the viewer might be that of human beings as exceptionally destined for good things. The catch of this particular science fiction brand of self-congratulatory ethnocentrism, however, is that these great achievements of humankind have, in fact, not yet happened.

Second, one must ask what the Prime Directive's and the series' distinction between "pre-warp" and "warp" civilizations means when it comes to the predetermination of historical development and change. Are all civilizations, left to their own affairs, destined to steadily evolve into that of Starfleet and the Federation? Are these to be understood as the pinnacle of creation?

An especially well-suited example of the development of civilizations in *Star Trek: Voyager* is given in the episode "The Blink of an Eye" (season 6, episode 12). Here, the starship gets stuck in the atmosphere of a planet where time passes exceptionally fast. "For each second that passes on *Voyager*, nearly a day goes by on the planet," we are informed. First officer Chakotay is excited by the unique possibilities of studies this entails, while his colleague, chief engineer and half-Klingon, half-human B'Elanna Torres, is jaded:

> **Chakotay:** This could be the greatest anthropological find of my career. If there's an intelligent species down there, we'll be able to track their development not just for days or weeks, but for centuries.

Torres: Watch them discover new and better ways of beating each other over the head.
Chakotay: They won't necessarily follow the Klingon model.
Torres: As opposed to the human model?! It will take a few hours to make the adjustments [to the ship's equipment].
Chakotay: A few hours. We might miss the rise and fall of a civilization.
Torres: So we'll watch the next one.

As their banter indicates, if there is actually a human and a Klingon model of development, both species have a violent history (although Klingons are even in the series' present well-known for their honor-based culture's emphasis on physical strength, fighting, and drinking). The ship's involuntary position, stuck in the discovered planet's atmosphere, does however let its crew follow the development of a civilization, and it is a rather familiar development.

As an audience, we are treated to a few glimpses of this development, from ancient times, when *Voyager*'s presence as a light in the sky, causing earthquakes on the planet, is interpreted as a godly presence; to a space-age civilization. This development is tracked by Chakotay and Torres and seems to hold no great surprises, although it is starting to become clear to them that the presence of their own ship is somehow affecting the culture they are watching.

Torres: The next series of scans are coming through. I'm downloading them into the display buffer.
Chakotay: No doubt about it. There's a city down there.
Torres: Elevated levels of carbon monoxide, ammonium—that's progress, alright.
Chakotay: They've developed internal combustion technology since the last few scans. Look at these radial lines. It looks like a system of roads.
[---]
Chakotay: Look at the amount of iron being used in that city. That's ten times what you'd expect in a culture at this stage of development.
Torres: If you lived on a planet that wouldn't stop shaking, you might be doing the same thing. If they reached this stage of development, they must be observing us.

Here, it is clear that the development on the planet follows a predictable path where cities and cars and steel-reinforced buildings lead to space observation and space travel—underlined by a cut after the last quoted line to precisely a space observatory on the planet, where scientists are indeed observing *Voyager*. The catch is, the presence of the alien spaceship, causing constant earthquakes, causes the planet's inhabitants to focus on developing weapons technology to blow the disturbance out of the sky. It actually takes a violation of the Prime Directive, since the planet's population is not yet warp-capable, to meet them and persuade them not to blow the ship up, but instead to help *Voyager* leave the planet's atmosphere. Without the presence of *Voyager*, the impression is that the planet in "Blink of an Eye" might have completely followed the same developmental curve as Earth, supporting the impression of Earth's development as something of a norm in *Star Trek: Voyager*.

There are, however, plenty of civilizations that might or might not have developed warp-technology according to this curve, but still give a somewhat different impression than the people of Starfleet and the Federation. I have already mentioned the Klingons, often a subject of cultural and physical clashes with humans in the Star Trek franchise, but alien encounters are frequent. Since *Star Trek: Voyager* takes place in a different part of space from the other series, we are introduced to plenty of new species, as well as some familiar ones. In the upcoming parts of this chapter, I will argue these encountered alien species are not only to be perceived as Others but can actually be read as a metaphor of the human past of Star Trek—our own past or even present. As opposed to the apparent Othering of the alien species encountered by *Voyager*, I argue they can in fact be seen as representations of ourselves and our past. We were Ferengi, we were Malon, and we were even a little bit Hirogen.

We Were Ferengi

The Ferengi are an alien species who appear throughout the Star Trek franchise, from *The Next Generation* on. They are depicted as greedy, scheming, and untrustworthy, and at the same time audacious and comically cowardly, often causing trouble for the Federation main characters. When given more recurring roles, like bar owner Quark and his family on the space station Deep Space Nine in the Star Trek series of that name, even Ferengi are shown to have more complex personalities.

In *Star Trek: Voyager*, Ferengi appear in only one episode, called "False Profits" (season 3, episode 5). In the beginning of this episode, the *Voyager* crew discovers the planet Takar. Their instruments are "picking up an M-Class planet. Humanoid life signs. However, metallurgical analysis indicates a pre-industrial civilization, a Bronze age level of technology." At the same time, they are "detecting evidence that these people have had contact with the Alpha Quadrant," specifically "a modulated energy discharge that appears to be consistent with the use of a replicator," a Federation piece of technology that produces food or objects seemingly out of thin air.

After Tovuk and Chakotay go down to the planet to investigate, they discover that a couple of Ferengi have made their way to the Delta Quadrant through an unstable wormhole in space. On Takar, they have completely taken control of the population, appearing to them as gods or prophets known as "the Holy Sages." The two Ferengi are accumulating great wealth, while devastating the planets population. Simultaneously, they are feeding them their own values, the capitalist wisdom of the Ferengi scripture *The Rules of Acquisition*, with catch phrases such as "Greed is eternal" and "Exploitation begins at home."

When Chakotay and Tuvok return to *Voyager*, their condemnation of the Ferengi is quite strong.

> **Chakotay:** It seems the people have a myth, an epic poem called "Song of the Sages," which predicts the arrival of two demigods from the sky, the Sages, who would rule over the people as benevolent protectors.
> **Tuvok:** But these Ferengi are anything but benevolent.
> **Chakotay:** What they have done is co-opt the local mythology by using advanced technology like the replicator to convince the people that they're the two Sages spoken of in the poem.
> **Tuvok:** Of course, being Ferengi, they haven't just co-opted the mythology. They've cornered the market. On everything.
> **Chakotay:** It's disgusting, Captain. The two Ferengi live in a palatial temple, while the people are lucky to have a roof over their heads.
> **Tuvok:** Apparently, it wasn't always like that. According to the people that we met, before the Ferengi came, the society may have been primitive, but it was flourishing.

Although the Federation principle is to not interfere with other cultures, Captain Janeway finds a loophole in the fact that the Federation can be seen as partially responsible for the Ferengi making their way through the wormhole and decides to intervene and send the Ferengi back where they came from. "I certainly don't intend to leave them here to continue exploiting an innocent society," she states. In the end, the *Voyager* crew manages to send the Ferengi away in a way that harmonizes with the local myths of the Sages, leaving the people of Takar satisfied and free of exploitation.

The behavior of the Ferengi in this episode is depicted and described as completely despicable. It is quite clear that the *Voyager* crew would never sink to such behavior. However, to the viewer, the actions of the Ferengi might just as likely seem familiar. A great deal of their behavior is strongly reminiscent of the colonial era in Earth's history. The Ferengi have arrived in a new land, where they learn only enough about the natives' way of life to exploit the local beliefs to their own benefit. They ruthlessly gather riches, relying on their superior technology and enforcing their own belief system on the local population.

The lifestyle of the Ferengi is that of extreme capitalism. Greed is their culture and religion, but theirs is also a society of great inequality. In the episode "False Profits," no female Ferengi are depicted, but the male pair of them interact mainly with other men, while keeping young, beautiful, and scantily clad local women around for massaging their lobes (the preferred erotic pleasure of the big-eared Ferengi). The Ferengi male is a patriarch, and, as we've learned, "Exploitation begins at home."

Neither greed nor capitalism nor gender inequality are strangers to human history (or for that matter, to the present). With the Ferengi, we are free to despise and laugh at these qualities as something other to our own characteristics, as Otherness. At a closer look, however, might the creature in the distorting mirror of science fiction appear familiar? One could argue that the profit- and status-seeking Ferengi, constantly counting their gold-pressed latinum, are more similar to today's human beings than the idealized, non-capitalist citizens of the Federation. Imagining a humankind of the future, free from greed, might be the most radical of all of Star Trek's futuristic visions.

We Were Malon

The Ferengi are far from the only profit-hungry aliens in the Star Trek universe. In the episode "Night" (season 5, episode 1), *Voyager* reaches a starless void, heavy with theta radiation. Before they know it, they are under attack, but another vessel chases the attackers away. When the captain of this other vessel greets them, it is with the words "I had to fire thirteen spatial charges to drive those ships off. I expect to be compensated."

If somewhat brusque, the alien captain, a Malon named Emck, first appears friendly. Then, the *Voyager* crew encounters one of the beings that attacked them in the first place. This creature is indigenous to the Void and dying of radiation poisoning. So are, in fact, all of his species. The Malon are dumping their toxic waste in their habitat. "They are poisoning our space. We don't know why," the alien explains. "We tell them they are killing us. They won't listen. We tried to stop them. They are too strong."

With this insight, Captain Janeway and Chakotay confront the Malon captain:

Janeway: You're using their space as a dumping ground for your anti-matter waste. Why?
Emck: My civilization produces six billion isotons of industrial by-product every day. This region is the perfect disposal site.
Chakotay: How convenient. For you. A spacial vortex in the middle of nowhere, far away from your own system. Out of sight, out of mind. The problem is, somebody lives here.
Emck: *One* species.
Janeway: One's enough. We didn't come here to debate the issue. We came here to offer a peaceful solution.
Emck: What kind of solution?
Janeway: My people use anti-matter as well, but we've found ways to purify the reactant so there's no toxic waste. We'll show you.

The Malon is shown the technology to purify and recycle his toxic waste. He is impressed. But he turns the offer down, saying: "This would solve a lot of problems on my world. Unfortunately, it would also put me out of business."

The Starfleet crew cannot believe what they are hearing, and Chakotay tries to argue against the Malon's fear of being made obsolete and throwing "the waste export industry into chaos."

Chakotay: We're proposing changes, some of them difficult. But progress can also bring new opportunities. Given time, this could turn to your advantage.
Emck: I already have the advantage—the vortex. No one knows about it, except me and my crew. By ejecting my cargo here, I cut expenses in half. I won't sacrifice that.
Torres: I guess mass murder doesn't factor into your profit margin?!

Not being able to persuade the Malon to stop dumping his toxic waste, *Voyager* manages to destroy the vortex, and thereby close off the waste hauler's profitable shortcut into the Void.

Despite technical advancements similar to that of the Federation, the Malon choose a different path than them. Rather than dealing with their waste problems, they chose to dump them on other, defenseless creatures, far away from their own home planet. "Out of sight, out of mind." This strategy is, of course, not unlike those of Earth's industrial era, dumping toxic waste in the ground, the ocean, in the wilderness, or simply shipping it off to be dealt with by less fortunate people in other parts of the world. This analogy may not be explicitly pronounced, but it is certainly there for interpretation. We were, and are still, Malon.

We Were Hirogen

The *Voyager* crew's first encounter with the alien species Hirogen occurs in an episode called "Message in a Bottle" (season 4, episode 14). In this, *Voyager* establishes contact with Starfleet for the first time since arriving in the Delta Quadrant. They do so using alien technology; a network of relays they believe to be abandoned. However, the network is claimed by a Hirogen, who will not let them use it any further. Without the captain clearing her approach, Seven of Nine sedates the uncooperative Hirogen by using the network connection to give him an electric shock. She justifies this strategy simply with "He wasn't responding to diplomacy."

This description returns as *Voyager* continues to cross paths with the Hirogen. In the episode "Prey" (season 4, episode 16), Chakotay concludes: "From what I've found in their database, diplomacy isn't a part of their lifestyle. They don't see us as equals. To them, we are simply game." Even though Captain Janeway is determined to prove them wrong by showing them compassion, the Hirogen respond at best with doubt and contempt. The *Voyager* senior staff is also skeptical, with Seven of Nine openly defying the captain's orders, and the usually logical Tuvok passing judgment on the Hirogen with the assessment "I believe we should consider them extremely dangerous. They seem to lack any moral center."

Tuvok and Seven could be excused. In the episode "Hunters" (season 4, episode 15), they are attacked in their space shuttle and taken captive by a pair of Hirogen, credited only as Alpha-Hirogen and Beta-Hirogen. The prisoners wake up in what looks a bit like a torture chamber, filled with instruments and trophies like skeletons and skulls. The Hirogen even complain about what an easy catch they were: "You were pathetic prey. Easily taken. The hunt was not satisfying." Still, they are happy with their rare find, since their parts will make original trophies. "Unusual relics are prized," one of them explains to their prey. "Yours will make me envied by men and pursued by women."

In the following episode, "Prey," *Voyager* takes in a wounded Hirogen, and thereby gets access to his ship and its database. This allows them to analyze the Hirogen more carefully.

> **Chakotay:** The entire culture seems to be based on the hunt. Social rituals, art, religious beliefs. They're nomadic. Their existence is driven by the pursuit of prey, and it's carried them across huge distances.
> **Tuvok:** There is no evidence of a home planet. Their ships travel alone or in small groups. On occasion, several will join forces in a multi-pronged attack.
> **Chakotay:** Like wolves.

To the *Voyager* crew, this lifestyle is quite strange. They even exchange a look of curious disgust at the fact that the Hirogen seem to, in part, use their victims for food. This is something the species of the Federation have put long past them, gaining their nutrition from replicators or, when the opportunity presents itself, from the vegetable kingdom.

Unlike the Federation, having put hunting and meat eating behind them with technological (and moral) development, the Hirogen, despite having made obviously similar technical advancement, have not. Below, we will return to their ability to "evolve," but first, I wish to point to our similar origins. Humans were, and are, of course, hunters. But this is not the only dark side of humanity's past (and perhaps, present) the Hirogen episodes explore: as we shall see, this also includes fascism.

Klingons Versus Nazis

The Hirogen are soon to return to *Voyager*. In the two-part episode "The Killing Game" (season 4, episodes 18 and 19), they have taken over the ship, and, with the help of neural implants, reduced the crew to characters in a great and deadly holodeck game. They base these games on information from *Voyager*'s database on Federation history. As the Hirogen leader, the Alpha, states: "These people have a violent history." As the Doctor is one of few *Voyager* crewmen to maintain his memories, being left to patch up all who are wounded in the holodeck simulation, he is able to inform Seven of Nine about the situation: "This has been going on for nineteen days. Dozens of battle scenarios, one more brutal than the next. You should see what a mess you were after the Crusades …"

The episodes do not, however, start with human history, but *in media res* with Captain Janeway as a Klingon, fighting a Hirogen. The Hirogen is pleased with this "resilient prey," deals her life-threatening wound, and then has her sent to sickbay to be patched up and reprogrammed to join the majority of the crew in a simulation of Second World War France. The smaller simulation of the Klingons warriors, drinking before a battle, is, however, kept running, with a new *Voyager* crew member being tested as a Klingon.

While his Hirogen subordinates are anxious to get to the kill, the Alpha who has taken over *Voyager* has a specific purpose—to learn from his prey, and thereby evolve his species:

> **Alpha:** Each prey exposes us to another way of life and makes us re-evaluate our own. Have you considered our future? What will become of us when we have hunted this territory to exhaustion?
> **Beta:** We will travel to another part of space, search for new prey, as we have always done.
> **Alpha:** A way of life that hasn't changed for a thousand years.
> **Beta:** Why should it?
> **Alpha:** Species that don't change—die. We've lost our way. We've allowed our predatory instincts to dominate us. We disperse ourselves throughout the quadrant, sending ships in all directions. We've become a solitary race, isolated. We've spread ourselves too thin. We're no longer a culture. We have no identity. In another thousand years, no one will remember the name Hirogen. Our people must come back together, combine forces, rebuild our civilization.
> **Beta:** What of the hunt?
> **Alpha:** The hunt will always continue. But in a new way. I intend to transform this ship into a vast simulation, populated with a varied and endless supply of prey. In time, this technology can be duplicated for other Hirogen. These holodecks will allow us to hold on to our past, while we face the future.

This conversation takes place with the pair of Hirogen dressed in full Nazi uniforms, since the simulation they are currently playing is that of World War II. The holodeck scene is that of a small town in France, occupied by

Nazi Germany, but on the verge of being liberated by the Americans. In this simulation, *Voyager* crew members play the part of the French resistance movement, with Captain Janeway as its leader, while Chakotay is the Captain leading the American troops. The reasons for the Hirogen to choose the losing side of the war are never made clear, although they seem to slip in and out of their Nazi characters with a natural ease. There are degrees to this, however. The Beta is clearly drawn to the Nazi ideal of strength, blood, and superiority. The more visionary Alpha, on the other hand, seems to find the rhetoric quite tiresome and unfounded. At one point, he snaps at a subordinate human Nazi officer, telling him: "You yourself, are you stronger than these degenerate races? More cunning? And if you were alone, without an army supporting you, would you continue the hunt? If your prey were armed instead of defenseless, what then? You are superior to no one!"

There is of course a certain comical effect to the Nazis going on about being a superior race, while failing to notice that their leader is actually a bumpy headed, redish-skinned alien. The Nazis seem, in fact, to lack the ability to learn and change that the Hirogen Alpha calls for. He finds these qualities in the Allied Forces from the World War II simulations, a resistance soon intertwined with the resistance of the *Voyager* crew against the Hirogen take-over of their ship. In negotiations with Captain Janeway, the Hirogen Alpha credits humanity with being resilient and cunning. "You seem," he says, "to recognize the need for change."

These ambitions will, however, be the death of the Alpha, as his subordinate Beta turns against him, inspired by the Nazi rhetoric. His reign does not last long, however, and a third Hirogen leader can call a ceasefire and leave *Voyager*. As the Hirogen withdraw, Captain Janeway bestows them with the technology needed to create their own holodeck. The new Alpha will make no promises of using it, saying how he is not as "unconventional" as his predecessor.

In my opinion, the conflict between, in part, what the original Hirogen Alpha describes in "The Killing Game" and what is represented by his assassin, the Beta, and the Nazi occupational force with which he is so closely linked, constitutes something essential in the Star Trek universe. Starfleet, though not perfect, is portrayed as striving toward change, openness, exploration, and learning. The Hirogen Alpha's ideas of learning from other people touches on this ideal, but distorts it somewhat by still referring to those encountered as "prey." The cyborg enemies of the Federation, the Borg, have a similar unscrupulous approach, as they accumulate knowledge from everyone they meet, but do this by "assimilating" them, thereby effectively killing or brainwashing them. In the final battle against the Hirogen occupying *Voyager*, the ship's crew (both in their capacity as themselves and as World War II resistance and the Allied Forces) is joined by the Klingons from the other holodeck simulation. The Hirogen and Nazis are therefore defeated by interstellar diversity, very much in contrast to the inflexible, hierarchical Nazi/ Hirogen way of thinking. In fact, one can argue, they are beaten by the multicultural diversity that is Starfleet—and Star Trek.

"And Now, for the Conclusion"

At first glance, science fiction may appear as the opposite of history. It is fantasy, imagination, freed from the ties of the present and past. But is it, really? Can we think about the future without taking some sort of guidance from the past? The frequent use of historical references in *Stark Trek: Voyager* certainly seems to imply that we cannot.

In this chapter, I combine John Fiske's concept "multitude of voices" with that of historical culture. A popular text does not speak in one, conclusive voice. The text can never be taken to mean only one thing—if it did, it could not be popular. The voices put together in the text, Fiske claims, come from those available in the surrounding culture. This is true, also, of the historical culture. When, for example, grasping for a historical image of evil arrogance and cruelty, Nazi German is an easily recognizable point of reference. Had the Hirogen episodes of *Voyager* instead chosen for example those of the Rwandan genocide, the images would risk failing to carry meaning to a wider (primarily Western) audience. As a TV series is not only a product of historical culture but also produces historical meaning it does, however, also risk reproducing the same commonly known images of history, never really challenging the status quo.

This is true not only of iconic historical events, persons, and phenomenon but of ideas of historical change and development. In several examples discussed in this chapter, *Voyager* perpetuates strong ideas of human (or alien) development as a forward, teleological movement, one of progress toward ever greater technical and moral achievements. At the same time, these ideas are rather vaguely phrased. The openness of the text makes it more prone to evoke questions than to give unambiguous answers.

In this chapter, I have focused mainly on what might be called a metaphorical use of history in *Star Trek: Voyager*. While it is completely possible to interpret alien species in Stark Trek as Others, as opposed to the mainly white and Western, Earth-centered Starfleet, I argue for the possibility of another interpretation, that of them as an othering of ourselves. We were (to the future), we are (in the present) the greedy Ferengi, exploiting those less fortunate than us. We are the unimaginative Malon captain, dumping toxic waste where it does not immediately affect his own environment, uninterested in a solution unless it gives direct profits to himself. We even were, and perhaps still are, the Nazi Hirogen, too comfortable at the top of the food chain to want to rethink and change his ways. By representing our own past or present as Other, concurrently putting our past, present, and future alongside each other, we can see them from a distance that might prove illustrative and constructive. If the future Starfleet could put these problems behind them, then maybe so can we?

References

Andres, Katharina. 2013. "Fashion's Final Frontier": The Correlation of Gender Roles in Star Trek. *Culture Unbound* 5: 639–649.

Fiske, John. 2001. *Television Culture*. London: Routledge.

Grech, Victor. 2020. The Banality of Evil in the Occupation of Star Trek's Bajor. *Early Human Development* 145: 105016.

Heath, K.M., and A.S. Carlisle. 2020. *The Voyages of Star Trek: A Mirror of American Society through Time*. London: Rowman & Littlefield.

O'Connor, Mike. 2012. Liberals in Space: The 1960s Politics of Star Trek. *The Sixties* 5 (2): 185–203.

Orwell, George. 1949. *1984*. London: Penguin.

Ott, Brian L., and Eric Aoki. 2001. Popular Imagination and Identity Politics: Reading the Future in Star Trek: The Next Generation. *Western Journal of Communications* 65 (4): 392–415.

Star Trek: Enterprise. 2001–2005. Paramount.

Star Trek: Voyager. 1995–2001. Paramount.

PART III

Defining and Defying the Boundaries of Cultures and the Human

CHAPTER 10

The Wild Boar Never Strikes Without Cause: Monstrous Hybrids, National Identity, and Gender in the Horror Movie *Chawu*

Anna Höglund

The Transcultural and Hybrid Character of *Chawu*

The wild boar has found its own unique space in the genre *animal horror*. The subgenre is minor and consists of only a few films. The Australian *Razorback* (1984) is the most influential, and it has reached iconic status as one of the most acclaimed films in the genre of animal horror. The subgenre of wild boar horror films may be limited, but it is consistent in it themes and motifs. It always describes conflicts between nature and culture, animal and human, and the criticism of civilization is a prominent theme. Often the exploration of the place, both the nature and the culture, which produced the monstrous boar, takes up a lot of space in wild boar horror. The landscapes where the wild boar causes destruction are liminal spaces between past and present, and the genre describes monstrous places and sites of conflict where something has gone horribly wrong. The South Korean film *Chawu* (2009) is no exception to the rule.[1]

A. Höglund (✉)
Linnaeus University, Småland, Sweden
e-mail: anna.hoglund@lnu.se

J. L. Hennessey (ed.), *History and Speculative Fiction*,
https://doi.org/10.1007/978-3-031-42235-5_10

Chawu opens dramatically by immediately establishing the main conflict in the film between human and animal and the theme of criticism of civilization. Strong scenes with animals being barbarically tortured are accompanied by newsreaders stating that "animals are distressed." In the following scenes in the film, the theme is further reinforced through a presentation of the life conditions in the capital city Seoul and that in the sparsely populated countryside, which are both described as being in a state of moral degeneration.

Seoul is portrayed as a chaotic and dark place. Violence, drug abuse, and corruption are rife. The law is just a veneer that is not observed by those in power. It is a society in decay that is described with the words "rotten life in Seoul."

In the countryside, the deteriorated quality of life is confirmed through a presentation of the village's incompetent police force and the men in power who abuse their positions. Unbridled capitalism rules and the nearby mountains have had the trees ruthlessly removed to make way for golf courses and weekend farming lots to attract rich city people. Soon, however, the villagers will have something else to think about than economic gain. A monstrous beast is moving in the shadows of the exploited mountain regions, and it is hungry. Humanity has upset the ecological balance, and punishment comes in the form of a gigantic wild boar with a taste for human flesh.

From the opening scenes of *Chawu*, it is already clear that the film has a dystopian tone. *Chawu* describes a society that went wrong long before the ravenous monster makes its entrance and the fictional setting reflects a social realm where pseudoscientific beliefs about the law of nature are predominant. Life is dictated by maxims like "eat or be eaten" or "kill or be killed." The theme of victims and perpetrators is prominent in the film. For the majority of the characters, life seems to be a constant battle where they either hit or get hit, symbolically and literally. In *Chawu*, people are immersed in the constant reproduction of meaningless repression that only creates suffering, helplessness, and the desire for revenge. It seems to be impossible to break the vicious circle. In this context the title *Chawu* is apt because it pinpoints the film's main theme and primary symbolism. In the Chungcheong dialect, "chawu" is the word for "trap." The title is multifaceted and saturated with symbolic significance. The word *chawu* can refer to the tangible traps that are constructed in order to catch the wild boar in the film. But it can also be interpreted in a more metaphorical manner. *Chawu* is a tale about traps, both literally and figuratively speaking, and above all it is an exhaustive social commentary to the sentiment of being trapped in the discourse of existence. This is a menace that concerns both animals and human beings. The wild boar is trapped long before he is captured. Humans have destroyed his home, and he has no choice but to leave the mountains and search for food in the nearby villages. The human characters are also stuck in traps, mainly consisting of traditional ideological norms and values, created by their ancestors but sustained and reproduced by themselves. They are tormented by the burden of the past and the notion that their destiny is predetermined in a way that makes it hard for them to evolve and make progress when this requires a considerable change. It seems like they are caught "in between" because they can neither move backward nor forward. This has grave consequences for the characters, not least for their construction of self-identity.

The Korean concept of *haan* can be used to describe this phenomenon. *Haan* is a unique Korean concept that is believed to be a result of Korea's dramatic history (Jung 2011, 7). The country can be called a monstrous place in the sense that its history has been one of recurring repression and violent conflicts. Korea has been invaded by nations such as China, Mongolia, and Japan, subjected to long periods of colonial oppression, and literally torn into two by a bloody civil war.

Haan has been compared to the American concept of blues. It is described as a collective feeling of shared suffering and destiny, which is important for the creation of a South Korean identity.[2] In my interpretation, *haan* simultaneously expresses frustration and sorrow over the very difficulty of defining and constructing a specific South Korean identity, since the country's culture is shaped by the influence of several different cultures. In that sense, *haan* expresses ideas that both confirm and question the idea of a distinctive national character.

In South Korea, the term *mugukjeok*, "lacking in or having no nationality," is frequently used to describe the nation's identity (Ibid., 17). According to Sun Jung (2011), the concept is essential in studies of South Korean popular culture, since it perfectly captures its transcultural and hybrid character. South Korean popular culture can be described as odorless, but it "is not only influenced by odourless global elements, but also by traditional (national) elements" (Ibid., 3).

In popular culture *mugukjeok* is an internationally successful concept, not least in film. A hybridization of Korean film tradition and American blockbusters appeals to both a domestic and a Western audience, while simultaneously offering something different, something alien. *Chawu* is an excellent example of ***mugukjeok***. It is a hybrid shaped by influences from Asian and Western genre film.

The Western influence is particularly clear in *Chawu*'s many intertextual references to the genre of animal horror, especially the wild boar film *Razorback* (1984) and the shark film *Jaws* (1975). These films permeate the plot of *Chawu*, its characters, motifs, and themes. Several well-known scenes in the films are imitated and parodied, together with a few occasional scenes from the creature feature *Predator* (1987).

Chawu is also influenced by a number of Asian horror film traditions. A clear influence comes from ***kaiju eiga***, a Japanese genre in which gigantic monsters cause destruction. *Kaiju eiga* is, by tradition, highly critical of society, often portraying problems of an ecological, social, and political kind.[3] Another Japanese tradition is ***kaidan eiga***, the ghost story film. Korea's indigenous version is called ***kuei dam***, and it often involves a vindictive female ghost, ***wonhon***. In *Chawu*, a "madwoman" in the village takes on the appearance and characteristics of a *wonhon*. With her black, uncombed hair and white face, she frightens the men of the village, disturbs the patriarchal order, and problematizes ideas about maternity and social exclusion (Lee 1988, 23–25).

Korean horror is often described as a fusion of several different genres such as comedy, speculative fiction, and melodrama (Peirse and Martin 1988, 6). Melodrama is particularly significant for the narrative tradition that dominates contemporary South Korean film. In the South Korean horror genre, the melodramatic elements often mean that the horror is neglected in favor of a study of the fictional evil and of the characters who confront it (Byrne 2014, 188). In a self-reflexive spirit, a film can investigate a society that gives rise to evil, and this often leads to a greater understanding and empathy for the monster.

In *Chawu*, the wild boar, like the Korean people, is a victim of the country's dark history. It is a genetic hybrid, created by the Japanese army in order to finance its warfare. Indigenous wild boars have been crossed with foreign ones, resulting in gigantic, aggressive monster boars that devastate the indigenous nature. The mutant boar was created for war, death, and suffering, and it is suggested that this resulted in the creation of a man-eating species called *Horochoros Minor Chageni*. The boar is a product of the colonial past and a creation made by both man and nature. In this way, the monster is also permeated with ***haan***.

Although the film's monster is a victim, a being for which one can feel empathy, it plays an important part as the antagonist of the film. In the humans' fight against the hybrid boar, the film explores ideas about South Korean identity, in general, and about gender, in particular. A particularly burning question is: How do different forms of hybridization, in the past and in the present, affect South Korean identity creation and the construction, reconstruction and deconstruction of masculinities?

The Hybridization of South Korean Traditional Hegemonic Masculinity in *Chawu*

According to Sun Jung, the transcultural and hybrid character of popular culture has contributed to a hybridization of South Korean traditional hegemonic masculinity (2011, 3–4). In *Chawu*, the portrayal of the male characters shows an ambivalence toward both traditional and hybridized forms of masculinity. Hegemonic features are simultaneously praised and criticized. In the encounter with the film's monsters, however, there is an evaluation of the characters' respective masculine identities.

Four men set out to hunt the wild boar: the hunters Chun Ill-man and Baek Man-bae, the detective Shin, and the police officer Kim. The men share the goal of killing the monster, but they are driven by different motives. They also represent different forms of masculinity.

In the film the hunter Chun and the detective Shin represent stereotyped South Korean forms of traditional masculinity. Chun is an older man whose masculinity was shaped in a patriarchal tradition. He grew up in the village, and over the course of his life as a skilled hunter, he has acquired an obvious authority. In the Confucian tradition, Chun's masculinity is primarily characterized by the concept of *wu*, physical and martial strength, but it also has elements of *wen*, mental strength, and civilized control. In this way, he can be said to be an ideal man.[4]

Detective Shin is called in to lead the investigation of the maimed bodies left by the ravages of the wild boar. He is a polite and intellectual man who radiates an inner calm. Shin shows a clear contempt for weapons and violence. Shin represents what Sung calls *seonbi* masculinity, where mental skills, *wen*, are more valued than *wu*. Seonbi masculinity has high status in South Korea and is considered to represent virtues such as courtesy, fidelity, loyalty, integrity, and "cultural-scholarly attainment" (Jung 2011, 26–29).

Baek Man-bae is a professional hunter who is hired to kill the wild boar. He had been an apprentice to Chun but left the village to pursue an international career. Baek has the whole world as his hunting ground and is known from television and newspapers. In his professional life, he has achieved economic success and high status, competences that typify the construction of hegemonic patriarchal masculinity, in both the East and the West.

Police Officer Kim is young and portrayed as being lost in his role as a man. He has adopted a traditional patriarchal role as breadwinner and takes his work as a police officer in Seoul very seriously, but it gives him no satisfaction. Kim's authority as the head of the family is non-existent, and his home is in permanent chaos. Kim feels powerless both at work and at home.[5] He has no control over his life and is portrayed as a spineless victim of fate.

In *Chawu*, the more traditionally patriarchal men are initially portrayed in a positive light. Chun is the first to realize that it is a wild animal that is threatening the village. He takes the threat seriously and does everything he can to protect the villagers, while simultaneously showing respect to animals and nature.

Shin, the detective, is also portrayed as a man with a sense of responsibility. With his *seonbi* authority, he gets the police force to do their work more efficiently. Shin does not accept any abuse of power, and when he adopts a strict attitude toward disciplining the village police captain Yoo (who is the film's representative of excessive military violence and blind faith in authority), it reinforces his image as a man of principle, loyal to his *seonbi* ideal.[6]

The hybrid masculinity of Baek, the hunter, is at first portrayed in a more critical light. Baek may be a good hunter, but he has lost his respect for life, with a ruthless side that is revealed when he lets the villagers serve as bait in a trap for the wild boar. Baek's monstrous sides seem to be a product of his hybridization. He has been schooled by Chun, but has rejected his teacher's holistic and humble attitude to life in favor of a narcissistic Americanized lifestyle. This choice has serious consequences. Baek surrounds himself with exclusive status symbols. His hunting equipment is expensive and high-tech. He lacks an "authentic" moral compass, however, and a genuine social context in which to ground his self-esteem and his masculinity. Baek is disillusioned, a man who has lost his way and his goals in a postmodern world where there seem to be infinite possibilities.

In the final conflict with the wild boar, however, the expected masculine order is turned upside down. Baek assumes the role of a classical Hollywood hero and sacrifices his life to save the others. Chun and Shin end up with their halos askew. Both are shown to have defects in their character.

Chun drinks too much, and he sleeps off the intoxication at the time when his grandchild falls victim to the wild boar. As protector of the family, he has not done his duty and he thinks he has failed as a man. Chun is afflicted by *hen*, a mixture of profound hate, self-contempt, and failure, which can only be relieved by taking vengeance on the perpetrator.[7] Chun fails to realize his revenge, however, and he is unable to restore his male honor.

Shin's *seonbi* masculinity turns out to only be a veneer. Shin is in fact a kleptomaniac and lacks self-control.[8] His knowledge of Confucianism is shallow, and instead he creates his own Confucius-like proverbs when he feels they are called for. In the final battle with the wild boar, Shin is the one who most clearly betrays his ideal of masculinity when he exploits his rank to save his own skin.

When the battle is over, it is clear that neither Chun, Shin, nor Baek are ideal men. They have all failed in one sense or another. In the hunt of the wild boar, the monster has ended up in the background and the battle between the men in the foreground. Instead of listening to each other and cooperating, Chun and Baek choose to be rivals and mark their territories. Shin is too cowardly to contribute anything constructive. The men's hunt is thus unsuccessful. They fail to kill the monster and to confirm their position as ideal men.

That the men are trapped in a vicious circle of destructive behavior, which only generates gender stereotypes and identities of a pathological nature, becomes particularly clear in the very last minutes of the film. It turns out that Baek has survived the attack of the wild boar but has suffered an even worse fate. He has been captured by the village's madwoman and vicarious *wonhon*. Like the *wonhon* of the horror genre, the madwoman challenges the patriarchal order, and instead of tenderly looking after Baek, she puts a diaper on him, hangs him up on meat hooks from the ceiling and threatens to torture him if he does not call her "mother." The scene is full of symbolism. From a gender perspective, it can be interpreted as an ironic but horrific exposé of society's perverted gender roles. It can also be interpreted as an image of South Korea's colonial past, when the people were forced to obey a series of self-appointed fathers and mothers of the country. Or perhaps it is South Korea herself that is today's despotic mother? A country that, despite its independence, is unable to create a new, healthier society, but which just continues reproducing evil and suffering.

Chawu ends in the same spirit as it began, with the theme of victim and perpetrator, and the final scene is permeated with the concept of *haan*. The film's critique of civilization, however, is not entirely pessimistic; there is hope. This is not presented in the form of a man but of a woman. After Chun, Shin and Baek have fled the field, it is left to the police officer Kim to kill the monster. He is not alone, however. At his side is a young woman, Soo-ryun, who has been reluctantly allowed to come along on the hunt by the male hunters if she can "cook and provide entertainment."

The film audience is introduced to Soo-ryun early in the film, in a scene where she is reading a book written by her hero and ideal role model, Jane Goodall. Suddenly she hears screaming from a man. She runs toward him and in horror he points at another book lying in the grass. The book has a bloody severed hand clinging to it. They have found the remains of one of the wild boar's first victims. In contrast to the man, Soo-ryun is not easily startled. On the contrary, the finding excites her, because it means that wild boars are near and she can start collecting the data she needs for her research. This behavior is typical for Soo-ryun. She is a young woman with both feet sturdily grounded in the soil but at the same time well-oriented in advanced technological science.

Soo-ryun and her male colleague are camping near the mountain. In a modest tent filled with the latest technical equipment, they conduct research on how ecological changes affect the region's stock of wild boar. Soo-ryun is intelligent and resolute. She knows the indigenous nature and culture well, but she also has an open mind for global influences. Like Chun, she has a holistic world view, in which culture and nature, humans and animals are not regarded as separate entities, but as a connected whole. Soo-ryun is not interested in the male hunters' pathetic cock fighting. She is only concentrated on the mission to find the wild boar. When the male characters quarrel, she makes real progress in their common goal to track their prey.

Soo-ryun does not want to kill the boar. Her intention is to learn more about it and unlike the male hunters she has no weapon; instead, she uses her camera. When she is filming she makes the comments in English, a language that she masters, unlike the men in her company. This emphasizes her ambition to make an international career, roam the world, and "make real difference," just as Jane Goodall did.

The hunting expedition turns out to be a hard challenge for Soo-ryun. She tries to keep a straight face and maintain a neutral attitude to the hunters' ambitions. But when they find the wild boar's den and discover a litter of piglets, she can't control her feelings. The hunters decide to kill all the piglets to put an end to the wild boars' existence. Soo-ryun protests are fierce but she only manages to save one piglet because she persuades the hunters to use it as bait for their traps. When the hunters kill the litter she cries but after that she puts the surviving piglet in her backpack and takes good care of it. This action proves to be essential to how the film ends.

After Chun, Shin and Baek had left the hunt in disgrace. Soo-ryun and the police officer Kim manage to lure the wild boar into an abandoned factory. Despite her respect for all life, it is Soo-ryun who finally captures the wild boar with the aid of a drop-trap in the form of an elevator. She thus combines knowledge of the ancient hunting tradition with modern technology to create something new which brings the desired result.

Kim also shows great courage in the last battle with the wild boar. He is aroused from his passiveness and risks his own life to save his family and the villagers from the monster. Like the other men, however, he does not fully measure up. When Soo-ryun captures the wild boar it does not die at once but is in tormented by its wounds. Kim then takes advantage of the situation and gets Soo-ryun to promise to say that it was he who killed the monster if he is to put the animal out of its pain. Soo-ryun snorts in contempt but as is customary for her character she instantly puts the issue behind her and concentrates at more important tasks at hand, in this case the little piglet in her backpack. The piglet was used as bait in the elevator but saved by Soo-ryun after the trap crushed its father.

Soo-ryun keeps her promise, and in the closing scenes of the film, it is fairly clear that Kim is also caught in the trap of patriarchal reproduction, for he has no moral qualms about letting himself be hailed as the man who killed the beast. When the male characters enjoy their fifteen minutes of fame and compete for media attention Soo-ryun keeps her distance, as usual. Her own mission is completed and she knows something that she is bound to keep for herself. In one of the film's last scenes, though, the secret is revealed for the audience. At a beautiful flourishing mountain field, a cute sprightly piglet is seen munching on green grass. When the camera moves closer the piglet glares into the lens and smirks in a malicious way. True to her ideological convictions, where animals and humans, nature and culture, function as a connected whole, Soo-ryun has set the wild boar's offspring free. Free to thrive and decide its own destiny.

Summing Up

Chawu is a South Korean product. It is transcultural and hybridized. Its theme is pervaded by *haan,* and it explores issues connected to South Korea's national identity and the construction of gender, in the past and the present. The phenomenon of hybridization is examined throughout the film.

In *Chawu,* it is clear that hybrids can be monstrous. The film's monster, the wild boar, is a hybrid. South Korea as a nation is also a hybrid, shaped under the influence of other countries' culture. Since the country's past is characterized by conflicts and violence, the nation and its history are also presented as monstrous. The past is a ghost that haunts the South Korean people and creates *haan.*

It is obvious that South Korea, in some sense, must free itself from its past. As the film's title and primary theme suggest, the South Korean people are trapped in the discourse of existence. This affects both the country's nature and culture in a negative way. In particular, it has a devastating influence on individuals' attempts to construct a satisfying self-identity. This article primarily illuminates two aspects of identity construction: the longing for and aspirations to mold a proud collective national identity and the struggle against oppressive traditional gender roles. This concerns both women and men. The characters Chun, Shin, Baek, and Kim are all caught in gender role traps. They are trying to form their masculinities by obeying the rules and ideals constructed by a hegemonic patriarchal tradition. Despite their efforts they are not happy and it is clear that they feel uncomfortable in the stereotypical gender roles they are trying to maintain. In the end of the movie, there is no hope for the men. They are still trapped. When it comes to Soo-ryung, it is a different story. She seems to be the only character, besides the surviving piglet, that has a promising future. Soo-ryung has found a way to avoid being caught in the trap consisting of traditional ideological norms and values. She does not discard wisdom from the past; instead, she makes use of it by combining traditional and contemporary knowledge. By setting the piglet free, she rejects the maxim "kill or be killed" and shows that she has no interest in claiming revenge on the offspring of the wild boar. It is clear that she refuses to take part in the reproduction of meaningless violence and repression that only creates suffering, helplessness, and the desire for revenge.

In spite of *Chawu's* harsh social critique of the conditions in South Korea, it is clear for the audience that nothing in the film's portrayal is either completely black or white. The nation's monstrous past also has precious knowledge that ought to be preserved. Even the film's monster symbolizes good values. The hybrid wild boar may have run amok, but he is also described as a protector of nature and animals. The wild boar defends the ancient mountain regions from human exploitation, so that the mountains can remain unchanged.

In *Chawu*, the important thing does not seem to be the origin of knowledge, but rather people's ability to employ it selectively, thereby breaking destructive patterns and preventing new ones from being established. *Chawu* possibly advocates a kind of principle of goodness, where the means and the ends minimize suffering and aim for the "good" for all living things. As history tell us, though, that is a principle that may well turn out to be a slippery slope. In *Chawu*, however, the will to at least try is portrayed as the only hope for South Korea and, by extension, for all of humanity.

Notes

1. The original South Korean title of the film is *Chawu* but it is also known as *Chaw* (world-wide title) and *Chawz* (DVD title in USA). The words chawu and chaw is pronounced "chow" which means, "trap" in Chungcheong dialect. Chawu 2009.
2. "It is the ethos of groups or racial mourning. Many years of social injustice, political oppression, economic exploitation, or foreign invasion create the collective unconscious *haan* or *blues*." Min 2003, 6–9.
3. Korea has produced films of its own in the genre, such as *Yonggary* (1967) and *The Host* (2006). Richards 2010, 29.
4. For more information about the concept of *wu* and *wen*, see Louie 2002, 17–18.
5. "South Korean hegemonic masculinity embodies patriarchal authoritarianism where men were traditionally considered to be heads of family and the main providers." Jung 2011, 26.
6. Scholar Moon Seung-Sook argues that military service constitutes hegemonic masculinity in South Korea (2002, 89). Sun Jung uses the concept *violent masculinity* to describe the phenomena. Violent masculinity is defined by violence, authoritarianism, blind obedience, and physical hardship. Jung 2011, 26.
7. The term *hen* is a Chinese concept that can be described as a more original form of the concept *haan*.
8. Self-control is important in the construction of hegemonic masculinities in many cultures.

References

Byrne, James. 2014. Wigs and Rings: Cross-Cultural Exchange in the South Korean and Japanese Horror Film. *Journal of Japanese & Korean Cinema* 6 (2): 184–201.

Chawu. 2009. Directed by Jeong-Won Shin. South Korea: Lotte Entertainment, Polygon Entertainment. Pre GM. DVD.

Jung, Sun. 2011. *Korean Masculinities and Transcultural Consumption*. Hong Kong: Hong Kong University Press.

Lee, Hyangjin. 1988. Family, Death and the Wonhon in Four Films of the 1960s. In *Korean Horror Cinema*, ed. Alison Peirse and Daniel Martin, 23–34. Edinburgh: Edinburgh University Press.

Louie, Kam. 2002. *Theorising Chinese Masculinity: Society and Gender in China*. Cambridge: Cambridge University Press.

Min, Eung-Jun. 2003. *Korean Film: History, Resistance, and Democratic Imagination*. Westport: Praeger.

Moon, Seung-Sook. 2002. The Production and Subversion of Hegemonic Masculinity: Reconfiguring Gender Hierarchy in Contemporary South Korea. In *Under Construction. The Gendering of Modernity, Class, and Consumption in the Republic of Korea*, ed. Laurell Kendall, 79–113. Honolulu: University of Hawaii Press.

Peirse, Alison, and Daniel Martin. 1988. *Korean Horror Cinema*. Edinburgh: Edinburgh University Press.

Richards, Andy. 2010. *Asian Horror*. Harpenden: Kamera Books.

CHAPTER 11

Heritaging and the Use of History in Margit Sandemo's *The Legend of the Ice People*

Cecilia Trenter

The Legend of the Ice People: An Introduction

> "A mixture of myth and legend interwoven with historical events, this is imaginative creation that involves the reader from the first page to the last."
> —*Historical Novels Review* (Oughton 2008)

This chapter explores the use of heritage and history in the popular romance series *The Legend of the Ice People* (1982–1989) by Norwegian-Swedish author Margit Sandemo. The epos is the 47-volume multigenerational saga of a family and one of the best-selling series of novels in Scandinavia. My study of this series is motivated not only by its scope and popularity, with some six million readers (Nilson 2015), but by the fact that the series' specific features are an outstanding example of how popular romance, with its core component based on the love story and erotic desire, makes use of both fantastic elements and references to the past in a variety of ways to create a multifaceted story world. The chapter is arranged in the following way: after a presentation of the series, I will introduce the

C. Trenter (✉)
Malmö University, Malmö, Sweden
e-mail: cecilia.trenter@mau.se

J. L. Hennessey (ed.), *History and Speculative Fiction*,
https://doi.org/10.1007/978-3-031-42235-5_11

recognizable in terms of mimesis, the uses of history (*historiebruk*) referred to as the remediation of heritage ("heritaging") and *concurrences* as a methodological approach. Thereafter, the chapter explores the series' use of historical points of view to investigate how history is integrated into the love story and its roles in the narrative formula. The closing section presents a glimpse of the comprehensive involvement of the readers in online fora and the ways they negotiate, confirm, and refurnish the world of the Ice People.

The Legend of the Ice People: The Plot

The full series consists of three parts: (1) *The Legend of the Ice People*, 47 books, set in the years 1581–1960; (2) *The Witch Master*, 15 books, set in 1699 and 1715–1746; and (3) *The Legend of the Realm of Light*, 20 books, set in 1746 and 1995–2080. This chapter will focus on the core part of the story world of the Ice People. The series, originally written in Swedish, has been translated into nine languages, including English from 2008. This chapter quotes the Swedish version of the novels, with my own translations. For the sake of convenience, references to individual novels in the following discussion use both the English and Swedish titles, and a complete list of the series' titles in both languages, along with the year of publication for the Swedish original, is included at the end.

The story in the first part revolves around a Norwegian family and mountain clan, called the Ice People, who also form their own race in an isolated valley in the Scandinavian Mountains (which is why they are known as the Ice People), sprung from the ancestor Tengel the Evil who, according to legend, sold his soul to the Devil to gain eternal life. The price was a curse; in each generation of his future relatives, there will be a person born with supernatural gifts to be used to perform deeds in the service of evil. The story starts in 1581, when a cursed outlaw, the outcast and marginalized Tengel of the Ice People who has inherited the name of the evil ancestor as well as his monstrous physique, strives to turn the evilness into goodness using the gifts that come with the curse, such as magical skills and healing powers, to help other people. The saga is about his descendants, especially the women, whose lives take place at the intersection of magical powers, passions, and the fight between good and evil. The stories take place during a 700-year period and recount a fantasy-driven narrative centering on the tension between the family's paranormal power and universal human everyday life in changing historical settings. The family tree grows wide and creates a complexity of unique fates in each generation, united by the crime and curse of the evil ancestor and the common traits of the family. Some members of the family have supernatural powers, while others have only inherited family traits such as love of animals and children. Each part in the series constitutes a complete adventure that is either directly related to the fight against the family curse or constitutes a side story with a separate adventure, often a horror story or mystery.

To desire and to let the desire form, even tame, the protagonist and the object of the protagonist's love defines the overall narrative in the series of the *Ice People*. The process of desire includes the awakening of the (hetero)sexuality of the female characters and their masculine counterparts, which is redeemed in their union. Whereas pornography focuses on the voyeuristic relation between the reader and the text, popular romance embodies closeness and identification in the sensations of lust, as well as fear from experiences of sexual violence that are also present within popular romance (McCann and Roach 2021). In Sandemo's saga, desire is often dramatized in the narrative trope of beauty and the beast. Maria Nilson (2015) discusses this trope of how the unspoiled but newly sexually awakened female saves, by means of arousal, the man who is physically wild and characterized by beastlike traits. The development and awakening shape the individual and the couple into a complete, human whole. Being, and becoming, human is quite central in Sandemo's epos. The journey to become fully human could be described as a coming-of-age story or as a subgenre of the classical *bildungsroman* which describes the moral growth of the protagonist from childhood to adulthood, in which a change of character is central. As in Barbara Cartland's historical novels, the baser sexuality of the hero and the virginity of the heroine is a frequent plot point in Sandemo's epos (Ficke 2021; Nilson 2015). Virginity is much easier to dramatize in the past than in the present, as Cartland once put it (Ficke 2021, 120).

The Recognizable, Heritaging, and Concurrences as a Methodological Approach

Images of the past are useful tools to arrange the main themes in the romantic story. The fictitious past is a popular trope within the romance genre, which frequently rejects realism by portraying fantastic elements such as time travel and extraordinary settings in exotic and/or historic contexts (Thurston 1987). Popular romance is reviled by historians for being shallow and riddled with errors, whether failing to properly reflect the historical period's *zeitgeist* or through inaccurate depictions of historical facts. For such critics, historical romance is simply a display of anachronism and arbitrariness in the use of history, and its historical settings are randomly picked and interchangeable with any exotic environments (Queckfeldt 2000). The historical settings are of minor relevance and subjected to the love story; the fact is that the average reader doesn't care about the details of the Battle of Hastings, as Leslie Weinger puts it in her handbook in historical romance (cited in Nilson 2015, 63).

Historical popular romance is not defined by historical accuracy, but instead generates a sense of authenticity through the expectations of the reader and genre conventions (Ehriander 2015; Nilson 2015). Historical settings in female-coded literature, such as popular romance and chick lit, actually aim to highlight current themes from the present. Or, as Helene Ehriander points out, the important themes the female confronts in the historical novel do not differ from those of Bridget Jones and her friends and are likewise recognizable for the female reader (Ehriander 2015).

The recognizable also lies in the references to cultural or historical expressions that Sandemo interleaves with the paranormal and fantastic. Although references to the past and an active use of history are central in Sandemo's series, the books are not historical novels. Margit Sandemo has no ambition to write history in the sense of teaching about or conveying Nordic history to the audience, even though the series comprises a multigenerational family saga that stretches over 700 years. In a chat with readers in the Swedish newspaper *Aftonbladet*, Sandemo describes her relationship with history-writing:

> **Reader:** How do you think when you put your main characters in other times and cultures? Do you do a lot of research?
>
> **Margit Sandemo:** No, I don't. I do the research afterward and it always turns out right. (Svärdscrona 2001)

In spite of the critique of how history is used in popular romance and Sandemo's categorical rejection of researching her historical milieux in advance, I argue that the relationship to the past is omnipresent and multidimensional through a mimetic use of the past. I use "mimetic" in relation to the aesthetic term for resembling or imitating nature with the aim of stimulating the viewer's imagination in sensuous and concrete ways. Based on Vivian Sobchack's term *conceptual mimesis*, I define mimetic elements as a general idea rather than likenesses of a historical person, event, or artifact (Sobchack 1990, 14). When Sandemo lets us meet historical persons, it is because they play a dramatic role in the romantic plot.

When the story's framework is built up on established literary models, it is because these contribute to staging the romantic plot. Although Sandemo's fictive stories are not intended to represent the past as a goal in itself, I argue that her use of history constitutes a cohesive narrative that is best described as a kind of mobilization of cultural heritage ("heritaging"). History and memory are selective. To create memory, it is not enough to refer to an isolated phenomenon or a historical fact. The phenomenon must be placed in a meaning-creating context in which both the contents (the historical references) and the form (style, genre, medium) transform the references to a memory. Literature has several similarities with memory processes. In research on collective memory and the use of cultural heritage and the past, emotions and imagination have been identified as driving forces in the active use of history. Awakening nostalgia, producing horror, and moving the reader through empathy or identification are recurring ways of emotionally relating to the past (see, for example, Smith et al. 2018). The focus of recent research has shifted from the cultural heritage itself to the performative use of cultural heritage, known as *heritaging* (Crouch 2010, 69). It is this form of "heritaging" that characterizes *The Legend of the Ice People*.

I argue that the historical settings are not interchangeable but play a crucial part in creating the family saga. Consistent with theories in memory studies that highlight remediation and dynamic heritage, I consider the use of history as a remediation of pre-existing, fully recognizable representations of the past. The use of the past is expressed in symbols, narratives, monuments, historical persons, commemoration days, symbolic actions, and geographic landscapes, and condensed into memory, shared by a collective, whether a local group, a nation, or in transnational communities (Erll and Rigney 2009). Astrid Erll and Anne Rigney argue that all historical representations are rearrangements of already existing heritage: from national monuments to historical representations in popular culture (2009). Fantasy fiction retells and reuses myths, folkloristic narratives, symbols, and heritage to create a credible universe (Höglund and Trenter 2021, 15).

Kathryn Hume states in her classic work (1984) that all literature consists of impulses of both fantasy and realism; the mimetic elements are crucial to embed the fantastic narratives. In other words, stories can be un-realistic but never incomprehensible. Sandemo highlighted the elements of *The Legend of the Ice People* that were based on real life. By her own account, she was herself psychic and the paranormal elements in her writing like spirits and demons in part reflect her own experiences (Sandemo 2011). The connection to reality also involved other experiences. Sandemo revealed near the end of her life that she was sexually abused in her childhood and that the descriptions of sexual violence in the series have a therapeutic function (Gilhus 2012). In the final books in the series, the elements of personal experience are further strengthened by Margit Sandemo herself appearing in the stories (*The Calm before the Storm/ Lugnet före stormen; Is There Anybody Out There/ Är det någon därute?)*. Sandemo strengthens the authenticity of the fiction by interleaving realism (mimesis) and the supernatural (fantasy) with real-life events, which has contributed to the attention the series has received (Gilhus 2012).

Sandemo's saga is defined as paranormal and fantasy, but several books in the series use typical gothic adaptions as a mode of literary expression. This includes a haunted house or castle and a dark secret that somehow drives the heroine or hero (Nilson 2015; Toscano 2021). The gothic hybridity in Sandemo's series revolves around secrets of the past and the family curse. Nilson (2015) emphasizes the prevalence of hybrids between subgenres in contemporary romance literature, for instance steampunk, which blends sci-fi with the Victorian era. Sandemo's epos contents of paranormal, fantasy, horror, and sci-fi elements, in particular the last of these in the closing parts of the series. I suggest that her universe could rightfully be called fantastic due to the hybridity of genres that characterizes so much fantastic fiction. The fantastic combines ideas and abstractions about the world into comprehensible yet unrealistic stories, and the mixture of familiar and unfamiliar elements produces a sense of wonder (Höglund and Trenter 2021).

A *story world* is a narrative that is constructed concurrently in several different media by multiple actors (Ryan and Thon 2014). In the case of Sandemo, this consists of the series of novels and fan fiction that takes place in and further develops the world of the Ice People. (*The Legend of the Ice People* has never had a film adaptation). In the following sections, I will first discuss the different dimensions of the uses of history before considering how the uses of history are negotiated by the readers. There is some existing research on fan cultures (see, for example, Duits et al. 2016), but in this chapter, fan cultures are approached from the perspective of the uses of history rather than focusing on them as cultural or social phenomenon in themselves. I argue for the importance of studying the story world of the Ice People from a methodologically holistic perspective. I take, as a point of departure, the necessity to interpret the complexity of the epos according to the concept of *concurrences*. Concurrences is an umbrella concept that in a variety of ways emanates from the statement that culture can only be understood by studying interlinkages as well as tensions and frictions between different approaches which are equally valuable (Brydon et al. 2017). I suggest that Sandemo's epos is more than a narrative: it is a dynamic story world in which readers and fans co-create the universe by discussing, interpreting, rejecting, or embracing characters and plots in blogs, fan sites, and podcasts (Harvey 2015). Brydon et al. (2017) define concurrences as a multifaceted tool to explore for instance the complexity and transcultural communication in which knowledge is produced. They argue that cultural communities and transcultural communications consist of epistemic communities and knowledge-producing groups, whereas I define the story world of Sandemo as a space for an epistemic community populated by readers and fans. In the rest of this chapter, I will take a closer look at the cultural heritage and the canon of historical knowledge that are used to make the stories of lust and love visible.

How the Frame Stories Are Created by Historical Settings

Consistent with the traditions of the dangerous lover in Western literature (Lutz 2006), Sandemo's stories about the doomed family contain mysteries of desire, death, and eroticism. The world of Sandemo is a dangerous place due to the structurally violent societies that inhabit it. People in Sandemo's saga live under constant threat in a dystopian world in which oppression by authorities, mostly the church, wealthy elites, or the representatives of the legal system, is omnipresent, often concretized in sexual abuse and rape. Different kinds of detentions are recurrent; the theme of imprisonment is performed according to the historical circumstances. This pervasive theme plays out in a variety of settings: a hidden underground storehouse (*Winter Storm/ Vinterstorm*), children in undergrounds mines (*Evil Legacy/ Det onda arvet*), abandoned barns (*Blood Feud/ Blodshämnd*), mental hospitals (*The Scandal/ Synden har lång svans*), and in POW and concentration camps (*Hidden Traces/ Små män kastar långa skuggor*). Being caught in social traps or physical captivity is often combined with sexual violence or abuse: the prurient church warden who hunts witches among young, beautiful women; soldiers who commit gang rape (*The Stepdaughter/ Avgrunden*); the priest who chastises and lusts after supposedly sinful women (*Devil's Ravine/ Vargtimmen)* and pedophiles who traumatize children (*The Knight/ Den siste riddaren*).

The background framing of this dystopian world is created through historical catastrophes. For example, epidemics are often featured, such as the plague in Trøndelag in Norway 1581, which causes chaos that forces the main character Silje to leave her previous life and become part of a new one, that of the Ice People (*Spellbound/ Trollbunden*). The historical context of social dissolution creates the framework for individuals' strong feelings and desires, both positive and negative, as when cholera affects Oslo in 1937 (*The City of Horror/ Stad i skräck*). In *Hunger*, the protagonist Marit is left behind when her siblings emigrate to America, leaving the ten-year-old girl to watch over her sick and evil father. She almost succumbs to sickness and hunger but is saved by a man of the Ice People, who also is physician.

However, war is by far the most important cause of structural violence that tears societies and individuals apart in the series. The brutality of war and the historical change that a war brings upon societies shape the characters and their moral status. The Thirty Years' War (1618–1648) forms an important part of the overarching plot to explain how the first two generations of the Ice People get separated. The war creates distance from the safe home of the Ice People back in Norway. The innate evil within the antagonist Trond develops during his experiences from the war, as does the innate goodness in his brother Tarjei, who is one of those chosen to fight evil. Rebellion in Norway against the Danish Kingdom makes up the backdrop in other books, like *Winter Storm/ Vinterstorm*. The Skåneland Wars (1675–1679) and the pro-Danish guerrilla "snapphanar" or "friskyttar" provide the background for the complicated relation between Villemo, who participates in the war cross-dressed, and Dominic, her beloved relative, who is captain over the soldiers that fight in the guerilla (*Yearning/ Feber i blodet)*. The protagonist Vendel in *The Eastwind/ Vinden från Öster* develops from naïve boy to man during his imprisonment in Russia as a prisoner of war in the early 1800s. The Second World War is yet another backdrop to dramatic happenings, including the Norwegian resistance movement (*Hidden Traces/ Små män kastar låga skuggor*).

The human is a collective creature in Sandemo's world. The individual, in Sandemo's series either she or he (the gender is thus binary), is only intact and perfected when he or she finds affinity with the opposite sex. True love, in contrast to false love, mere desire or shallow ideas about love, is more than a feeling between a couple; true love confirms the individuals' unity and "longing for association" (Reddy in Hsu-Ming 2021, 452). The worst a human can be exposed to is not war, poverty, or disease, but isolation. The theme of loneliness is a recurring one in *The Legend of the Ice People*, staged with the help of historical dramatization. It is, for instance, the family curse that keeps the witch Sol from feeling any earthly love for men. The beautiful, brave, and independent woman finds her own path in her search for the only lover who can satisfy her—the Devil himself, who appears in her fantasies—although always defending family members when they get in trouble. Sol responds to her loneliness by getting involved in the witch activities during the sixteenth century, hiding from the inquisition and joining the ecstatic orgies at "Blåkulla" with the Devil (*The Stepdaughter/ Avgrunden*; for similar themes about loneliness due to the curse, see *The Devil's Footprint/ Satans fotsteg*; *the Gardens of Death/ Dödens trädgård*; *The Dragon's Teeth/ Drakens tänder* and *The Ferryman/ Färjkarlen*). Sol, tired of running and hiding, finally surrenders to the authorities. The night before she is executed, she ingests deadly hallucinatory herbs:

> In the final hallucination, Satan appears, enveloping Sol with warm and understanding eyes, overflowing with true love. For the first time Sol could feel love and for the first time Sol was completely happy. (*The Stepdaughter/ Avgrunden*, 253)

Another protagonist, Mikael, is separated from his kindred, the Ice People, and becomes a loyal soldier and captain in the army of Swedish king Karl X Gustaf. His loneliness and depression grow deeper during war experiences in Livonia. His loneliness gets even worse when he marries a Catholic and virtuous woman in an arranged marriage. The long journey from being strangers to close friends and lovers is illustrated by the letters he writes from the wars during the 1650s. He experiences the Swedish king's march across the Öresund strait to capture Copenhagen and other campaigns that make up the background for his loneliness because of mental anguish and depression (*Without Roots/ Den ensamme*).

The loneliness of Sandemo's characters is often caused by the judgment of society, as in the case of a hangman's daughter who suffers from isolation and family stigma (*Under Suspicion/ Bödelns dotter*). The historical settings arrange the prerequisite for the loneliness; the hangman's stigmatized position during premodern times creates a convincing and explicit reason for his daughter's exclusion from society and illustrates the tolerance among the Ice People through the marriage and passion between one of the men of the Ice People and the hangman's daughter.

The Use of Time Markers

Sandemo's epos captures a wide span of historical epochs, from medieval times to after the Second World War, and the history of Scandinavia—which is the primarily geographic setting—is visible in different ways in the individual books. In the earlier part of the series, up until 1900, the historical canon in terms of named historical actors is displayed. The protagonist Cecilie in *Friendship/ Dödssynden* is governess for the children of Danish king Christian IV and his spouse Kristin Munk. Their real-life relatives Leonora Christina and Ellen Marsvin also appear in the stories of the Ice People. The protagonist of *Ice and Fire/ Is och eld* Viljar meets Marcus Thrane, a Norwegian activist who worked for the mobilization of the working class in Norway. Sandemo employs mimetic strategies in the form of temporal markers in order to create believability also in the story of the evil Tengel who contributes to the dystopian nature of the story world by working on the development of the atomic bomb and taking part in Nazi atrocities (*Hidden Traces/ Små män kastar långa skuggor*).

Sandemo also employs temporal markers through figurative language that evokes specific feelings connected to the historical references. For example, when Sol rides off into a stormy night: "A sonorous choir thundered in the crowns of the trees. It roared and mumbled in a deep tone like a choir of monks in a gigantic cathedral…" (*The Stepdaughter/ Avgrunden*, 4). Viljar, who lives in the nineteenth century, is described as "The great, dark stranger who mostly resembled one of the tragic knight figures of the Middle Ages, wavered in his self-imposed isolation" (*Ice and Fire/ Is och eld*, 62).

Historical markers are woven in as part of the novels' plot, not infrequently in concrete ways that are central to the development of the story, such as when the rebellious Elisabeth refuses to wear the then-fashionable Rococo powdered wig both because she feels stifled by the strict hairstyles and because she is allergic to the powder, which becomes decisive for the resolution of the adventure (*Behind the Façade/ Bakom fasaden*). The vain Euphrosyne kills a young man with a candelabra that turns out to have been silver-plated iron after he calls her "pasty-faced" (*The Mandrake/ Galgdockan*).

The View of History in Sandemo's Story World

Margit Sandemo's view of history is characterized by a focus on ordinary people at the bottom of the social ladder and a critique of the elite's abuse of power. Royal actors are not depicted in a positive light, and Sandemo consistently adheres to a view of the monarchy and other authorities as oppressors of the people. Protagonist Sol meets a boy on her journey who gives a detailed update about the latest conflict between Swedish royals and the nobility that results in the Linköping Bloodbath (1600). Sol recalls the death of Swedish king Gustav I Vasa, but is unaware of the following conflicts between his sons, a canonic part of Swedish history, well-known to the Swedish readers from schoolbooks. Upon hearing the news, Sol shrugs and answers "Well, power struggles there too. Like everywhere else" (*The Stepdaughter/ Avgrunden*, 73). By letting the reader take part in a rather detailed account of Swedish political history in the late sixteenth century, Sandemo gets the opportunity to underline the lack of importance that the internal fights of the nobility have for the people, both Sol and the readers. Sol doesn't mince the words when she judges the late Danish king Christian IV: "The limp, fat carcass they carried out, dead drunk as he was...Christian IV is surpassed in drunkenness only by his late father, Frederick II" (*The Stepdaughter/ Avgrunden*, 196). Sandemo's critique also extends to socialist elites. When Belinda is able to join a secret meeting for revolutionary workers, she listens but is unable to understand their speeches about universal suffrage and land for the rural proletariat. She certainly understands that she is a part of the propertyless class that the revolutionaries are talking about but she dislikes being seen as a victim and is unmoved by their political ideas (*Ice and Fire/ Is och eld*).

Sandemo has a normative approach to actors as well as social changes in history. On the cover of *The East Wind/ Vinden från Öster* (1984), Sandemo underlines the debt the royals have to the people as a result of their decisions: "When Karl XII decided to conquer Russia, he had no idea how much sorrow and misery he would cause." People who migrated to America during the great emigration from Scandinavia are depicted as traitors, and this view is entangled with the narrative when the aforementioned protagonist Marit—a ten-year-old girl—is left alone with her abusive father while her siblings "flee" the country to seek a better life in America (*Hunger*).

Later, in modern times, new authorities are likewise depicted as oppressors. Protagonist Malin stands up to a bully at the local municipal authority who plans to modernize her town by digging up the churchyard in which the Ice People rest. Malin's reactions to the imposing head of the office show how masculinity is connected to public political roles. Malin guesses that the bureaucratic man must have problems with his wife, and therefore acts "with desperate authority" at the office (*The Brothers/ Människodjuret*, 118). Malin criticizes the young clerk at the office, Per Volden, who oversees the building plan, and comments on his looks by diminishing of his masculinity: "That servile functionary grimace makes you look like a snotty miss" (*The Brothers/ Människodjuret*, 120). After visiting the graveyard and listening to the stories of the Ice People who are buried at the cemetery, Per changes his mind and argues for saving the graveyard as cultural history and memorials to people with unique destinies. Per and Malin fall in love and get married.

In Sandemo's world, genuine leaders are produced by the people themselves, not the elites. Nevertheless, organized democratic movements like the women's rights movement or unionization are not depicted in a favorable light. When the enlightened medical student and member of the Ice People André meets his love Mali, she is a feminist from the working class. She is described as moody and noisy, but her life path leads her to André, whose love causes her to tone down her political engagement and makes her a trusting and calmer wife and mother (*The Woman on the Beach/ Kvinnan på stranden*).

In *Ice and Fire/ Is och eld*, the warm and simple girl Belinda meets the cold and socially isolated son of the estate-owner, Viljar of the Ice People. He disappears at night on unknown missions, but his nocturnal projects are slowly revealed as Belinda warms up his chilly personality: he is active in the workers' revolutionary plans. Thanks to Belinda's good sense, Viljar realizes that his involvement in the worker's movement was unnecessary since the Ice People's estate already takes care of its workers and crofters. Sandemo's criticism of modern society's political organization is crystal clear in Belinda and Viljar's love story. The tyranny of elites over ordinary people is condemnable, but the solution is not through modern social engineering but through a conservative, traditional family ideal.

Remediation of Literary Genres

The different parts of the book series belong to different genres. By mimetically staging each romance narrative in a particular genre, the basic story of the *bildungsroman* can be dramatized from different perspectives. The storyline of Heike, who appears in five books, takes place in historical settings ranging from the late 1800s to mid-1900s and depicts social changes during the period. But the use of the past is also present in the remediation of literary genres that are represented in the story of Heike. *The Wings of the Raven* is a gothic novel, and the setting is a Bram Stoker-like Dracula milieu, which starts with two aristocrats who have fled from the French revolution and end up in a village located in the shadow of the "Castle of the Witch." The following book, *Devil's Ravine,* follows the Shakespearian structures of a comedy of mistaken identities. The historical setting emphasizes the subjugation of individuals by class structures and religion. In the following book, *The Demon and the Virgin*, Heike arrives at the estate of the Ice People in Norway, of which one of the farms is his rightful inheritance. The story is focused on lust and inhibitions and borrows themes from the classic story of beauty and the beast (Nilson 2015). The next novel is a ghost story in which Heike summons evil spirits and demons to protect his estate (*Rituals/ Våroffer*). The final part, in which Heike has a prominent role, features Dickensian depictions of destitute mine workers (*Deep in the Ground/ Djupt i jorden*).

The Fantastic and the Uses of History

The world in Sandemo's series is dark and dangerous. The curse caused by the evil Tengel to get eternal life in exchange for his descendants performing service to evil has similarities both with the myth of Doctor Faustus, who sells his soul to the devil in exchange for wisdom, and the idea of the original sin, inherited by humans over generations who are therefore in debt to God and subject to death. As in similar fantasy worlds, the criticism of established religions is prominent in favor of supernatural powers and alternative systems of belief (Feldt 2016). As Gilhus puts it, "Sandemo combines traditional Norwegian folk beliefs and New Age language" (Gilhus 2012, 63). Sandemo accentuates the authenticity of the existing paranormal phenomena such as ghosts and demons with superstitions in the historical settings. People's fear of werewolves and ghosts is abused by the authorities (*Under Suspicion/ Bödelns dotter*), just as people's longing for a faith is abused by practitioners of religion (*The City of Horror/ Stad i skräck*). However, magic skills and the supernatural do exist. Demons are real and play a central part as objects of attraction (for example, in *The Flute/ Vandring i mörkret*), and the fallen angel Lucifer has a crucial role in the overall narrative (*Lucifer's Love/ Lucifers kärlek*). Ghosts appear both as evil creatures (*Rituals/ Våroffer*) and as lovers (*Demon's Mountains/ Demonernas berg*).

The temporal aspects of Sandemo's series are integrated by the fact that the dead ancestors become guardian angels and help the Ice People to fight evil and Tengel's spirit, which grows stronger during the centuries. In this way, Sandemo maintains continuity as the good spirits guide and support their living relatives. The spirits move between the spiritual realm and the everyday life of the living, and thereby develop the timeline not only from 1581 when the series begins but even before, as spirits from the thirteenth century show up when the final battle is approaching.

Sandemo also makes use of time travel as a temporal device. The closer to present times the story goes, the more important flashbacks become to create connections to earlier generations of the Ice People. Tova, who starts as both physical beastly and evil, is born in 1936—rather late in series. During her character's development, she travels in time by hypnosis and enters evil witches in the past. By experiencing her hosts' loneliness in their marginalized positions during witch hunts and exclusion from their societies, she learns to act less selfishly and becomes more empathic when she returns to her own time (*Imprisoned by Time/ Fångad av tiden*; *A Glimpse of Tenderness/ En glimt av ömhet*).

Travel through time is activated by different heritages. Cursed artifacts and places that carry difficult memories (see, for example, *The Garden of Death/ Dödens trädgård*; *The Secret/ Huset i Eldafjord*) communicate with the living people. The heritage in terms of folklore, old stories and songs, plays a central role in a love affair. Christa's mystic lover Linde-Lou turns out to be a ghost, who appears in a folk song about a brutal murder in a mythical past. By doing research about the origin of the folk song and the original song text, Christa not only solves the crime that inspired the lyrics but also finds out that Linde-Lou is a much older relative of hers (*Troll Moon/ Trollmåne*). When Christa, many years later, becomes a widow, she meets Linde-Lou, who has now joined the army of spirits of Ice People, which prepares for the final fight with the resurrected evil Tengel (*Demon's Mountain/ Demonernas berg*). The erotic encounter between the now middle-aged woman and her dead uncle, who is still 18 years old, as he was when he died, creates a temporal and paranormal twist to the time travel theme and a taboo cross-generational attraction between a mature woman and a young man.

The Story World of the Ice People in Fan Media

There is not much research on Margit Sandemo's authorship or production except from a couple of master's theses and a chapter in Maria Nilson's overview of romance literature (Nilson 2015). However, there are plenty of fan sites, blogs, and podcasts that discuss Sandemo's story world in which readers name their favorite novels, clear up uncertainties, and exchange experiences from the universe of the Ice People. The negotiations and discussions have been ongoing in various media for several decades. The context, apart from the content of the various books, has changed over the years. When a blogger discussed the series in 2015, the epic saga was compared to *Game of Thrones, True Blood* and *Twilight* (Foliant 2015). When the books were recorded as audio books, someone commented how the high quality of the reader, Julia Duvenius, created better conditions for the experience of the *Ice People* compared to the experiences from the books, which are riddled with typographical errors (Jönsson 2020).

The historical milieux are discussed as well. One blogger expresses surprise that, despite university studies in history and archeology that should prompt scrutiny in her reading, she tends to uncritically accept Margit Sandemo's historical settings about sixteenth-century Norway, demonstrating the author's power to convincingly portray characters and milieux (Jönsson 2020). The blog presents the starting book in the series, and a commenter on the blog replies that she loves reading about the past "and some magic and witches just makes it more exciting" (Amanda 2020). The comments indicate that the readability and perceived authenticity doesn't depend on representations of the past but rather on how the depictions correlate with the fantastic and romantic plots.

Another reader expresses happiness over the detailed descriptions of historical costumes within the stories. One reader's preexisting interest in a certain period-typical style becomes actualized when the book series reaches the epoch: "Finally a lovely description of the fashion of the time,

I like that. Hair rolls with bread crusts and hatpins and tight dresses" (Fröken Anna Maria 2016).

But there are critical voices as well. One blogger analyzes the changing character of the series:

> As already said, in most of the books it is the women and their fate that fascinate me the most. Therefore, I wonder why their ways change so drastically with the passage of time? From the 16th century onwards, there are plenty of strong and headstrong women with a forward-thinking spirit, something I have really appreciated as I read—according to tradition, the women of that time should have been more restrained. But no, the subdued women only start to appear when the 18th century turns into the 19th and 20th centuries. Then every other woman is suddenly helpless, scared and depressed and has to meet a strong and sensible man who can instill in her confidence and make her understand that she is actually adorably beautiful. Gunilla Grip, Marit in *Hunger*, Agnete, Vinnie and Christa are all restrained by a parent or similar and are set in very sharp contrast to Sol, Villemo, Ingrid and Vinga who happily color the book pages in the earlier part of the series. One of the few strong women who appears in the modern part of the story is Mali, and then she instead gets the label "women's libber" and if not hates, at least dislikes men—before she meets André Brink, who makes her rethink. Why does it feel like the women in Margit Sandemo's story could be freer and stronger in the 17th century? Isn't it just as entertaining to write and read about liberated women nowadays—when women can be liberated in a way they couldn't then? (Eli 2010)

Sandemo's focus on tolerance as a virtue among the Ice People does exploit groups and people that somehow are objects of intolerance in society: poor people, people suffering from mental illness, and homosexuals. The perhaps most explicit example of queerness, which is depicted in negative terms, is the protagonist Alexander who meets a daughter of the Ice People, Cecilie. He turns out to be homosexual, which forms his and Cecilie's character development and their growing love relation. Alexander finally gets "cured" after finding out that he has been victim of abuse as a child by a homosexual servant. Alexander explains "There are good and bad people among us, just like with you. Our disposition gives us no excuse for evil actions. But most of us are ordinary, decent people" (*Friendship/ Dödssynden*, 167). The war provides a chance to restart, since he gets wounded and paralyzed from the waist down. In the care of Cecilie, Alexander's physical and mental "damages" are fixed. Throughout the series, the normative pattern in the stories depicts a young woman (or man) who grows up and is confirmed as human in a heterosexual relationship. The very idea of "tolerating" homosexuality is a kind of objectivity that only deepens the stigma of being homosexual (Walters 2014). The double-edged notion of "tolerance" raises discussions on fan sites:

> I reacted to the way the books dealt with homosexuality. Surely, she says there is nothing wrong with being gay, but the only good gay character ends up having a happy life with a woman, while those who are more genuinely gay are portrayed as less nice individuals. Could Margit Sandemo have squeezed even one gay love story into all those generations? But writers write to tell their stories, not to be politically correct, so if you want to read the story, you simply must ignore it. (Insidan IFokus 2007)

In the podcast *Isfolket*, an alternative story to the romance between Cecilie and Alexander is considered. Instead of making Alexander heterosexual, the story could have stayed at the first stage of their marriage when the couple decided to live in a sham marriage and live a non-monogamous relation (podcast *Sagan om isfolket Dödssynden* 2015). A comment on the podcast explains the problematic depictions of Alexander's sexuality with "I guess it's quite typical of the time, both when the action takes place and when the book is written." The commentator draws the conclusion that Alexander might have been bisexual (*Frågor till er inför Sagan om Isfolket-podden 5. Dödssynden!* 2015).

Closing Words

This chapter has explored how the mimetic use of heritage in Margit Sandemo's *The Legend of the Ice People* helps to fill in and create the dark world in which narratives centered on lust and desire play out. Heritage and the references to the past in the dystopian world of Sandemo is remediated by structural violence and by performing exclusion and oppression, which create the necessary conditions for romance and moral growth from childhood to adulthood. Sandemo's view of history can be said to be that of a "people's history" where the authentic people are pitted against the machinations of the ruling class in its different guises throughout history. The historical settings frame the narrative of being alone before becoming fully human by growing up and finding love. The historical settings are furthermore the starting point for highlighting how desire and love form and develop the protagonists. Time travel and various kinds of heritage connect the Ice People through generations to pinpoint the universal truth that individuals must fight the omnipresent evil despite where in the past or present she or he might be.

The complex epos of the *Ice People* is here defined as a story world in which Sandemo is the creator, but fans are actively involved in collective negotiations. Facilitated by the concept of concurrences that advocates a focus on parallel understandings, both as intertwining and tensions and frictions, this chapter shows that Sandemo's popular romance highlights values about gender and society which are cemented by repetition in different historical settings, mimetic impulses, and by fantastic elements. But the use of the past within the series also offers explanations and interpretations to readers, who are putting the old-fashion and conservative modes into up-to-date, concurrent, and contemporary understandings of morality and lust by emerging from the historical past within the series.

The Legend of the Ice People/ Sagan om isfolket (Helsingborg: Boknöje AB)

1. Spellbound (*Trollbunden* 1982)
2. Witch-hunt (*Häxjakten* 1982)
3. The Stepdaughter (*Avgrunden* 1982)
4. The Successor (*Längtan* 1982)
5. Friendship (*Dödssynden* 1982)
6. Evil Legacy (*Det onda arvet* 1982)
7. Nemesis (*Spökslottet* 1982)
8. Under Suspicion (*Bödelns dotter* 1982)
9. Without Roots (*Den ensamme* 1982)
10. Winter Storm (*Vinterstorm* 1983)
11. Blood Feud (*Blodshämnd* 1983)
12. Yearning (*Feber i blodet* 1983)
13. The Devil's Footprint (*Satans fotsteg* 1983)
14. The Knight (*Den siste riddaren* 1983)
15. The East Wind (*Vinden från Öster* 1984)
16. The Mandrake (*Galgdockan* 1984)
17. The Garden of Death (*Dödens trädgård* 1984)

18. Behind the Facade (*Bakom fasaden* 1984)
19. The Dragon's Teeth (*Drakens tänder* 1984)
20. The Wings of the Raven (*Korpens vingar* 1985)
21. Devil's Ravine (*Vargtimmen*1985)
22. The Demon and the Virgin (*Demonen och jungfrun* 1985)
23. Rituals (*Våroffer* 1985)
24. Deep in the Ground (*Djupt i jorden* 1985)
25. The Angel (*Ängel med dolda horn* 1985)
26. The Secret (*Huset i Eldafjord* 1986)
27. The Scandal (*Synden har lång svans* 1986)
28. Ice and Fire (*Is och eld* 1986)
29. Lucifer's Love (*Lucifers kärlek* 1986)
30. The Brothers (*Människodjuret* 1986)
31. The Ferryman (*Färjkarlen* 1987)
32. Hunger (*Hunger* 1987)
33. Demon of the Night (*Nattens demon* 1987)
34. The Woman on the Beach (*Kvinnan på stranden* 1987)
35. The Flute (*Vandring i mörkret* 1987)
36. Troll Moon (*Trollmåne* 1987)
37. The City of Horror (*Stad i skräck* 1988)
38. Hidden Traces (*Små män kastar långa skuggor* 1988)
39. Silent Voices (*Rop av stumma röster* 1988)
40. Imprisoned by Time (*Fångad av tiden* 1988)
41. Demon's Mountain (*Demonernas fjäll* 1988)
42. The Calm before the Storm (*Lugnet före stormen* 1989)
43. A Glimpse of Tenderness (*En glimt av ömhet* 1989)
44. An Evil Day (*Den onda dagen* 1989)
45. The Legend (*Legenden om Marco* 1989)
46. The Black Water (*Det svarta vattnet* 1989)
47. Is There Anybody Out There? (*Är det någon därute?* 1989)

References

Amanda. 05/08/2020. Comment on Jönsson, Sandra. 2020. "Trollbunden—boken som inleder sagan om isfolket." sandrajonsson.se. Accessed on 25/09/2022. https://sandrajonsson.se/trollbunden-boken-som-inleder-sagan-om-isfolket/.

Brydon, Diana, Peter Forsgren, and Gunlög Fur. 2017. What Reading for Concurrences Offers Postcolonial Studies. In *Concurrent Imaginaries, Postcolonial Worlds: Toward Revised Histories*, ed. Diana Brydon, Peter Forsgren, and Gunlög Fur. Leiden: Brill.
Crouch, David. 2010. The Perpetual Performance and Emergence of Heritage. In *Culture, Heritage and Representation: Perspectives on Visuality and the Past*, ed. Emma Waterson and Steve Watson. London: Routledge.
Dödssynden, Podcast Sagan om isfolket, 18/10/2015. Accessed on 25/09/2022. https://podtail.com/en/podcast/sagan-om-isfolket-podden/5-dodssynden/.
Duits, Linda, Koos Zwaan, and Stijn Reijnders. 2016. *The Ashgate Research Companion to Fan Cultures*. Surrey: Ashgate.
Ehriander, Helene. 2015. Chick Lit in Historical Settings by Frida Skybäck. *Journal of Popular Romance Studies* 5 (1): 1–12.
Eli. 06/07/2010. "Sagan om Isfolket—Margit Sandemo." Eli läser och skriver. Accessed on 25/09/2022 http://elilaserochskriver.se/sagan-om-isfolket-margit-sandemo/.
Erll, Astrid, and Anne Rigney. 2009. Introduction: Cultural Memory and its Dynamics. In *Mediation, Remediation, and the Dynamics of Cultural Memory*, ed. Astrid Erll and Anne Rigney, 1–15. Berlin: De Gruyter.
Feldt, Laura. 2016. Contemporary Fantasy Fiction and Representations of Religion: Playing with Reality, Myth and Magic in *His Dark Materials* and *Harry Potter*. *Journal of Contemporary Religion* 46 (4): 550–574. https://doi.org/10.1080/0048721X.2016.1212526.
Ficke, Sarah H. 2021. The Historical Romance. In *The Routledge Research Companion to Popular Romance Fiction*, ed. Jayashree Kamblé, Eric Murphy Selinger, and Teo Hsu-Ming, 118–141. London: Routledge.
Foliant. 23/06/2015. Netflix nästa stora serie. *Nöjesguiden*. Accessed on 25/09/2022. https://blogg.ng.se/foliant/2015/06/netflix-nasta-stora-serie.
Frågor till er inför Sagan om Isfolket-podden 5. Dödsynden! 15/10/2015. Accessed on 25/09/2022. http://shailina.se/Isfolket/viewtopic.php?f=35andt=8217.
Fröken Anna Maria. 23/02/2016. Comment on "Frågor inför Isfolketpodden—32 Hunger!" Sagan om Isfolket. Accessed on 25/09/2022. http://shailina.se/m/Isfolket/viewtopic.php?f=35andt=8286andhilit=32+hunger.
Gilhus, Ingvild S. 2012. Post-Secular Religion and the Therapeutic Turn: Three Norwegian Examples. *Scripta Instituti Donneriani Aboensis* 24: 62–75. https://doi.org/10.30674/scripta.67409.
Harvey, Colin B. 2015. *Fantastic Transmedia: Narrative, Play and Memory Across Science Fiction and Fantasy Storyworlds*. London: Palgrave Macmillan.

Höglund, Anna, and Cecilia Trenter. 2021. Introduction. In *The Enduring Fantastic: Essays on Imagination and Western Culture*, ed. Anna Höglund and Cecilia Trenter, 3–20. Jefferson, NC: McFarland and Company.
Hsu-Ming, Teo. 2021. Love and Romance Novels. In *The Routledge Research Companion to Popular Romance Fiction*, ed. Jayashree Kamblé, Eric Murphy Selinger, and Teo Hsu-Ming. London: Routledge.
Hume, Kathryn. 1984. *Fantasy and Mimesis: Responses to Reality in Western Literature*. London: Routledge.
Insidan, Infokus Lotta U, 9/03/2007. Accessed on 25/09/2022. https://insidan.ifokus.se/discussion/86452/isfolket!
Jönsson, Sandra. 2020. Trollbunden—boken som inleder sagan om isfolket. sandrajonsson.se. Accessed on 25/09/2022. https://sandrajonsson.se/trollbunden-boken-som-inleder-sagan-om-isfolket/.
Lutz, Deborah. 2006. *The Dangerous Lover: Gothic Villains, Byronism, and the Nineteenth-Century Seduction Narrative*. Columbus, OH: Ohio State University Press.
McCann, Hannah, and Catherine M. Roach. 2021. Sex and Sexuality. In *The Routledge Research Companion to Popular Romance Fiction*, ed. Jayashree Kamblé, Eric Murphy Selinger, and Teo Hsu-Ming, 411–428. London: Routledge.
Nilson, Maria. 2015. *Kärlek, passion och begär: Om romance*. Lund: BTJ Förlag.
Oughton, Ann. 2008. Spellbound: The Legend of the Ice People. Review in *History Novel Society* 45. Accessed on 22/01/2023. https://historicalnovelsociety.org/reviews/spellbound-the-legend-of-the-ice-people/.
Queckfeldt, Eva. 2000. Om kiosklitteratur med historiskt motiv. In *Makten över minnet: Historiekultur i förändring*, ed. Peter Aronsson, 91–106. Lund: Studentlitteratur.
Sandemo, Margit. 2011. *Livsglädje. En biografi*. Stockholm: Schibsted.
Smith, Laurajane, Margaret Wetherell, and Gary Campbell. 2018. *Emotion, Affective Practices, and the Past in the Present*. London: Routledge.
Sobchack, Vivian. 1990. "Surge and Splendor": A Phenomenology of the Hollywood Historical Epic. *Representations* 29: 24–49. https://doi.org/10.2307/2928417.
Svärdscrona, Lotta. 2001. Jag vet inte vad idétorka är. *Aftonbladet*. September 14.
Thurston, Carol. 1987. *The Romance Revolution: Erotic Novels for Women and the Quest for a New Sexual Identity*. Champaign, IL: University of Illinois Press.
Walters, Suzanna Danuta. 2014. *The Tolerance Trap. How God, Genes, and Good Intentions are Sabotaging Gay Equality*. New York City, New York: New York University Press.

CHAPTER 12

Shadowing the Brutality and Cruelty of Nature: On History and Human Nature in *Princess Mononoke*

Martin van der Linden

Introduction

> Some people think that Japanese treated nature very gently up to a certain period, that is, until toward the end of the war, and that it was during the postwar period of rapid economic growth that they grew cruel. From certain phenomena it looks that way, but I think we have always been cruel toward nature—Miyazaki Hayao, in conversation with Satō Tadao, 1997. (Miyazaki 2014, 56)

This quote is taken from a conversation between the Japanese movie director, acclaimed animator, and author Miyazaki Hayao and film critic Satō Tadao in the wake of the release of Miyazaki's 1997 movie *Princess Mononoke*.[1] The movie is an animated historical fantasy epic set in late Muromachi Japan. The story follows the Emishi prince Ashitaka and his involvement in a conflict between the *kami* inhabitants of an ancient forest and a human community of ironworkers bordering the forest, consuming

M. van der Linden (✉)
Linnaeus University, Småland, Sweden
e-mail: martin.vanderlinden@lnu.se

J. L. Hennessey (ed.), *History and Speculative Fiction*,
https://doi.org/10.1007/978-3-031-42235-5_12

its resources. Throughout the conversation, the topic of environmentalism and nature continues to come up, which is a common theme in Miyazaki's movies often discussed by scholars.[2] I think the above-quoted statement is crucial for understanding Miyazaki's overall view on environmental history and his notion that humanity, throughout history, has never been able to fully live in harmony with nature.

Miyazaki has claimed that in making this movie, he was interested in exploring three historical themes: human relationship with and view of nature, the historical influence of ironworks in Japanese history, and the fate of the Emishi (Miyazaki 2014, 47–50). Throughout this chapter, I will argue that *Princess Mononoke* (*Mononoke Hime*) can inform us of Miyazaki Hayao's often-discussed views on environmentalism and history, as well as his influence on notions of Japanese culture and identity. I will study the three different communities portrayed in the movie (the Emishi, Irowntown, and the Forest) and how Miyazaki these communities to give voice to different historical counter-narratives and problematize common notions of history and historiography.

A Short Summary of *Princess Mononoke*

The story of Princess Mononoke takes place during the Muromachi period (1338–1573) and follows the fictional conflict between the "old world" of nature and its spirits, and the "new world" of human ambition and rapid industrialization. The movie starts with a boar *kami*,[3] poisoned by an iron bullet, turned *tatarigami* ("cursed *kami*", a vengeful entity) attacking an Emishi village. Their young prince, Ashitaka, kills the tatarigami but by doing so gets cursed with a deadly wound. Ashitaka gets exiled and leaves the village to find a cure, knowing that he will not be able to return to the village. Ashitaka decides to travel to the West, where the boar *kami* came from.

Near the end of his journey, he encounters two communities in conflict. On one side is Lady Eboshi's Irontown at the borders of an ancient forest which is ruled by the mysterious and mighty being named Shishigami.[4] On the other side is a pack of wolf *kami*, led by Moro the She-wolf and her human daughter San (or Princess Mononoke). Lady Eboshi, wishing for her community to thrive, wants to expand the ironworks of Irontown into the forest. However, this is the Shishigami's domain, which San and the wolves want to protect.

Ashitaka is sympathetic to both causes; he feels sympathy for Irontown, for its population consists of marginalized people whom Lady Eboshi has given a home; yet he is also understanding of San and her wolf pack's wish to protect the forest and nature. In a rapid unfolding of events, the conflict becomes more serious as Lady Eboshi moves to slay the Shishigami at the request of the emperor, who had heard that the Shishigami's head would bestow immortality. The forest's forces are aided by an army of boar *kami* to attack the forces of both the ironworks and the army of a local warlord seeking to besiege Irontown.

The forest's forces lose the initial battle and Lady Eboshi succeeds in decapitating the Shishigami by way of blasting off its head with a musket, giving it to the emperor's servants who then flee the scene, for this act of defilement spells disaster for both sides. The Shishigami's corpse turns into a massive dark and headless giant (*Daidarabotchi*), searching for its head, destroying the land around it by killing everything it touches: humans, animals, and the entire forest. In the end, Ashitaka hunts down the emperor's servants and recovers the head, giving it back to the aimless Shishigami's corpse, saving the day, and healing his curse in the process. San and Ashitaka stay friends, but they cannot be together; San cannot face the humans who wished for her forest to be burnt down. Ashitaka goes to the ironworks, and San returns to the forest, thus ending the movie on a melancholic note. With the onslaught of the Shishigami stopped, the forest is revived, but the Shishigami remains dead. And although Ashitaka tells San that the Shishigami is still alive, San is unconvinced. Nature returns, but its spirit is killed and Ashitaka's mark will never fully disappear.

The Last of the Emishi

Although it is never actually specified in the movie itself,[5] Miyazaki in several interviews said that the story is set in the Muromachi period, but no specific dates or locations are provided (e.g., Miyazaki 2014, 16; 34; 60–61). And despite the ad hoc periodization and that the depiction of the Muromachi countryside is akin to the aesthetics of a Kurosawa Akira movie, such as *Seven Samurai* or *Rashomon,* the narrative design choices surrounding the three main communities of conflict in the movie, the Emishi village, Irontown, and the Forest of the Shishigami, are fictional and set in a fictional space and time, separate from any historical occurrence.

The first community shown in the movie is Ashitaka's, the hidden Emishi village. Historically, the Emishi were an ethnic group who lived in independent communities (*mura*) in the mountainous northern regions of Japan's main island of Honshu. The few historical sources left regarding the Emishi are biased accounts written by members of the Yamato court in the south of Honshu, describing the Emishi as being "Eastern savages" with primitive lifestyles, lacking both farming technology and proper housing. These descriptions are discredited by archaeological evidence indicating that, despite cultural and linguistic differences between the northern and southern regions, Yayoi agrarian culture had spread to the northern regions about two centuries after its inception in the south and the northeast and northern regions had a technology level comparable to the Yamato culture (Friday 1997, 3–4).

When the Yamato court in the south of Honshu started to garner influence and power during the eighth century, they started to extend their borders to the north, where they came into conflict with Emishi communities that had their farmsteads, fishing waters, and hunting grounds there. Starting in 774, the Yamato embarked on a military and colonial subjugation mission. Beyond periodic warfare, the Yamato court settled outposts in the northern regions, declared jurisdiction over Emishi land and communities, and recruited Emishi leaders to their cause, coopting their power (Friday 1997, 7). Lacking the military might of the south, the Emishi used guerilla tactics to combat the Yamato armies, resulting in many setbacks for the Yamato expansion. But in 811, the Yamato court declared the Emishi to be officially subjugated and the villages that had not participated in the resistance got assimilated into the Yamato cultural hegemony (Friday 1997, 23). This assimilation suggests the possibility that many Emishi communities survived after subjugation for a long time. However, the Muromachi period (1338–1573) is 500 years after the subjugation wars, and it is quite implausible for the Emishi village, as depicted in the movie, to have remained isolated and hidden for so long. Yet in Miyazaki's mind, small, independent villages remained. Hidden and isolated, their presence on the verge of extinction, the village and Ashitaka himself, embody one of the thematic leitmotifs of the movie: marginalization.

Ashitaka and his village are the last of the Emishi, their village lies hidden deep in the mountains of Honshu, and he is their last prince,[6] as the elders of the village note at the start of the movie. The curse from the wild boar carries the threat of almost certain death for Ashitaka and at the same time dooms the village to fade into obscurity. The Emishi shaman,

Hii-Sama, declares that Ashitaka's curse will mark him an outsider, and he must leave the village. Ashitaka cuts off his topknot (*mage*, the symbol of his manhood in the village), as a visual marker of his outsider status and leaves in the dark of night. Kaya, a girl from the village, against the village's laws and customs, meets Ashitaka at the entrance and gives him a small obsidian pendant in the form of a sword as a final gift to remember her (and the village) by. Ashitaka accepts this gift. This pendant is later in turn gifted to San at the end of the movie.

Virtually nothing has survived of Emishi culture, religion, customs, and art, and thus many of the materials on display in the movie are plausible historical fictional creations by Miyazaki and his art team, visually relating the movie's Emishi village to certain historical periods and cultures. In one of the first scenes of the movie, Ashitaka rushes to the village watchtower, which visually is reminiscent of the reconstruction of a great wooden tower found at the Sannai-Maruyama Iseki archaeological site in Aomori, Japan, dating to the Jōmon period (14,000–300 B.C.). The addition of this tower informs the audience that the Emishi are an ancient people, but not people foreign to Honshu because Aomori is in the northern part of Honshu, in the Tohoku region. Other visual cues, visually moving the movie's portrayal of the Emishi away from the dominant culture is villagers' dress, reminiscent of the traditional clothing worn by Ainu people in Hokkaidō, the northern island of Japan; clothes that are not used by other people in the movie. Here Miyazaki and his team refer to the widespread claim that the Emishi and Ainu people are related, yet modern scholarship has started to dispute this (Friday 1997; Matsumura and Dodo 2009; Hudson et al. 2017). Nevertheless, these active design choices, as Eija Niskanen points out, make it clear that "Miyazaki's version of the Emishi tribe is therefore visually and culturally distinctive even within *Princess Mononoke's* fantasy world" (Niskanen 2019: 45).

References to the isolated and hidden nature of the Emishi in the movie include the fact that Ashitaka never actually reveals his origins to anyone, despite people around him constantly asking him due to his steed Yakul (a fantasy creature called a "red elk" akin to a Lechwe from western Africa), the gold he carries and his clothing. The only person Ashitaka tells the location of the village is San, at the end of the movie as a sign of deep respect and trust between the two.

Despite the unlikelihood of Emishi culture to have survived into the Muromachi period, the movie makes sure to underline that the village has its own culture. By visually encoding the indigenous Emishi village as both indigenous to the land, as well as distinct from the dominant culture, as Niskanen observes, Miyazaki goes against the nationalist and nativist *Nihonjinron* ideology and historiography where the notion of "Japaneseness" is bound to certain visual cues not present in Miyazaki's portrayal of the Emishi (Niskanen 2019, 48). This narrative design choice enables the audience to understand that (1) the Emishi are an isolated group of people, hiding from the rest of Japanese society; yet, by using visual codes such as the Sannai-Maruyama Iseki tower and the "Ainu-esque" clothing; (2) the Emishi are not portrayed as outsiders in their own country; rather, they are depicted as a minority culture. Different from the dominant culture, but in no sense less connected to Japan.Their claim to identity is as strong as the other cultures portrayed in the movie. Miyazaki noted in an interview that: "Though they [the Emishi] have been obliterated, they were Japanese people, as it were. They had an independent state before Japan became unified" (Miyazaki 2014, 49). This marginalized nature interested Miyazaki, and despite the subdued role of the Emishi village in the movie as a whole, its spirit of resistance found within Ashitaka's character is one of the driving forces throughout the plot. Their marginalization, on the other hand, is symbolized by his curse mark.

The red–black bruise (*akaguroisaze*) of the curse mark grows bigger and bigger throughout the movie until it is foretold to devour and kill Ashitaka. And when the mark's power is used black phantom-like worms ooze out of Ashitaka's arm. But this mark also gifts Ashitaka monstrous strength, strength that can easily behead a man with little effort. But, as Benjamin Thevenin notes, while in other movies this mark and its powers would have been interpreted as a gift, in *Princess Mononoke* the audience can only interpret it as a curse, for it allows the otherwise peaceful and kind-hearted protagonist to commit horrendous acts of unnatural violence (Thevenin 2013, 162–163). The mark not only signifies the physical curse laid upon Ashitaka and how its power corrupts and changes him, it also signifies the end of the Emishi.

Commenting upon the nature Ashitaka's curse at the end of the movie, Miyazaki stated the following in a roundtable discussion moderated by manga scholar Makino Keiichu and partaken by philosopher Umehara Takeshi, Buddhist priest Kōsaka Seiryū, and medievalist Amino Yoshihiko:

> **Miyazaki**: His mark did grow fainter. Young people nowadays aren't convinced by a happy ending. They would feel it is more realistic not to have the

> mark disappear completely, and to have Ashitaka continue to live, bearing the burden of something that might flare up again at any time.
> **Umehara**: It is the mark of discrimination.
> **Miyazaki**: Yes it is.
> (Miyazaki 2014, 113–114)

Miyazaki's intentions are clear. Ashitaka and his mark, in the end, become a symbol for marginalized people, people who are shunned by common historiography, and even when their history is brought to the foreground, they must bear the mark of discrimination forever.

Miyazaki's stance on marginalized groups was part of a national debate in the 1990s and early 2000s regarding national identity, what it means to be Japanese and the question or rather criticism of Japanese homogeneity (Napier 2001, 474). Miyazaki shows an awareness of these issues in his movie and the portrayal of the Emishi reject the notion of Japan as an ethnically homogeneous nation. This is accentuated by setting the movie during the Muromachi period, an era often seen as quintessentially "Japanese," marked by a long period of civil war (*Sengoku jidai*, "warring states period," 1463–1603) that continued into the Azuchi-Momoyama period (1568–1600), when the power of the emperor had waned and the Shogunate fought for supremacy against other warlords and their samurai. This period is immortalized in thousands of books and movies; Miyazaki's contribution is to highlight that not even these periods, "the apex of Japanese high culture" (Napier 2005, 233), were as homogeneously ethnically "Japanese" as nativist narratives try to portray them. By making the story's hero a Emishi prince Miyazaki points out that the Japanese nation always has been, and always will be, a nation of cultural diversity and multiplicity.

The Utopia of the Ironworks

The subversion of nationalist historical narratives in *Princess Mononoke* continues when it comes to a major set piece in the movie. Miyazaki has said in an interview that the movie is set in the Muromachi period because it can be closely compared to the twentieth century; with its change in attitudes toward nature, social disorder, and humanity moving farther away from the supernatural toward an anthropocentric worldview (Napier 2005, 237). The second community, Irontown (*Tataraba*) is a very on-the-nose, powerful depiction of these changing times.

Lady Eboshi (*Eboshi Gozen*[7] in the original Japanese) is the leader of Irontown, an isolated wooden fortress with large ironworks at the edge of a great lake, bordering the Shishigami's Forest. Her town produces iron and metal weapons which it trades with merchants. The iron is produced by cutting down and burning the forests, so the town can build waterways to make the gathering of iron-sand easier. Angering the inhabitants of the forest. The town has no farmland of its own and thus must trade for food with neighboring towns. The settlement is populated primarily by former prostitutes, soldiers, homeless people, and iron workers, as well as a group of lepers whom Lady Eboshi gives relief in a small garden area of the town. The town is thriving, and its spirit of community is portrayed as strong and unbendingly loyal to Lady Eboshi and her cause to create a society where the dispossessed have a place and function. The women work the bellows of the *tatara* ("furnace") and work as hard as the men. The iron production and technology depicted in the movie are quite accurate according to archeological research (Kozuka 1968; cf. Inoue 2010). Irontowns's gender equality on the other hand is fiction. The main form of defense used by the townspeople is hand canons and muskets created by the ironworks' smiths. Ashitaka learns that it is bullets from one of these that hurt the boar *kami* at the beginning of the movie and turned him into a *tatarigami* and that it was Lady Eboshi who pulled the trigger. Lady Eboshi is thus also a military commander and a fearsome warrior in her own right, who does not fear the wrath of the *kami*, and someone whom the emperor personally commissions to kill the Shishigami.

With the inclusion of Irontown, Miyazaki once again goes against history and introduces a female leader as the main ambitious counter-cultural force of the movie. Certainly, the communities in the movie belonging to the dominant culture: a trading town Ashitaka visits before continuing his journey, which Irontown has dealings with, and the local warlord whose forces lay siege to Irontown at the end of the movie, all have the air of the Kurosawa-esque Sengoku era often seen in Japanese historical dramas, but Irontown is different. Susan J. Napier has pointed out that the movie portrays a historically utopian society by featuring two strong women, San (Princess Mononoke) and Lady Eboshi as the leaders of two conflicting counter-cultural communities (Napier 2001, 480–481). One can argue that the inclusion of Irontown, like the Emishi village, is an attempt by Miyazaki to go against the grain of contemporary and widely held historiography.

Women had no position of power in the Muromachi period. During this period, established laws of property rights and inheritance declined, and women could no longer inherit from either their fathers or their husbands (Ackroyd 1959, 43–44; cf. Kurushima 2004). Institutions were put in place to reduce the influence women could gain, and a marriage law was instituted that would make the bride the property of her husband and his family (Ackroyd 1959, 47–48; cf. Tonomura 1990). Likewise, despite ironworks and metal manufacturing growing more common throughout the country (Hall and Takeshi 1977; cf. Kenichi 1980, 14–27), as Napier points out: "the notion of such a community being led by a woman who is also a military commander and fiercely determined fighter is clearly fictional" (Napier 2001, 482). The wife of an influential lord, merchant, or another form of a high-status man in Muromachi Japan would have been able to enter certain spheres of political, labor, and economic influence, via their husband (Goto 2006, 195). But Lady Eboshi is unmarried, and throughout the movie, it is made clear that she has no partner or plans to get one; therefore, her position of power is fictional.

Miyazaki's movies are famous for their portrayal of powerful women. From the female military leaders in *Nausicaä of the Valley of the Wind* (1984, "Kaze no Tani no Naushika") to the ambitious and entrepreneurial Kiki in *Kiki's Delivery Service* (1989, "*Majo no Takkyūbin*"), it is unsurprising that the main characters in a position of leadership in *Princess Mononoke* are female. Yet this contradicts all available historical evidence about the status of women during the period. Therefore, the isolation of Irontown is very interesting; the secluded nature of the town, comparable with the hidden nature of the Emishi village, makes it a plausibility for such a community to have existed, despite historical evidence of the opposite. A hidden Emishi village, inhabited by a people long forgotten and suppressed, and a secluded ironworks, headed by a female warrior and populated by prostitutes and lepers, become historical possibilities. Furthermore, the isolated and dispossessed nature of Irontown thematically expresses Lady Eboshi's ambitions of leadership and agency, ambitions that women in the Muromachi period would not have been able to pursue.

In the abovementioned interview with Satō Tadao and elsewhere, Miyazaki has spoken more in depth about the characterization of Lady Eboshi as a "twentieth century person" put in the Muromachi period Japan. This creates a contrast with her surroundings that is remarked upon even within the movie. Lady Eboshi is strong in the face of adversity and always has her eyes on her target. As Miyazaki recounts in an interview:

> I like Lady Eboshi as a character. She's the epitome of a twentieth century person. To people from earlier periods, modern people must seem like the devil. They clearly distinguish between their aims and means and will use any means to reach their goals, while at the same time holding on to something pure. That is like the devil, isn't it? The devil is so dashing in the way he strides about. (Miyazaki 2014, 57)

Lady Eboshi is visually coded to be exceptional: she is tall, dressed in a blue silk kimono, self-assured "like the devil" and a character of action. Contrastingly, unlike the forces of the forest or the hidden Emishi, Lady Eboshi's isolation is self-imposed. Strengthening her claim to social independence beyond the norm. Her characterizing strength and presence is found in her cause and her belief that what she is doing is right, even if it means angering the *kami* themselves. Furthermore, it should be noted that when the monk Jigo, a servant to the emperor visits Irontown to deliver a request from the emperor to kill the Shishigami, she carries it out, not to help the emperor but to secure the position of Irontown and to save it from the neighboring warlords that throughout the movie have tried to besiege the fortress.

She finished the job of beheading the Shishigami by blasting it off with a shot from her musket. Moro the wolf *kami* had been beheaded just before the Shishigami's death, but her disembodied wolf head moves one last time and bites off Lady Eboshi's arm. The defiling killing of the Shishigami cannot remain unpunished; something Lady Eboshi reflects upon at the end of the movie. At the end of the movie, it is said that Lady Eboshi was carried by a wolf back to Irontown's lake; however, this is not depicted in any scene.

Lady Eboshi's character is the complex embodiment of human ambition and instinct to survive by whatever means possible. Many of her actions throughout the movie might in the eyes of some audiences mark her as the villain of the story, but as the narrative comes along, such a conclusion proves to be hasty. For example, like many of her actions throughout the movie, the shooting of the boar *kami* Nago was motivated by her desire to protect the Irontown community. Her hubris and ambition make her act in ways that might have dire consequences, but within the heat of the moment might seem rational. This characterization of human hubris is already present in her introductory scene. We see her for the first time in a narrow and slippery mountain pass with her ox drivers and soldiers, trying to transport goods back to Irontown while heavy rain is falling. This is the opportunity Moro, San, and their wolf pack take to attack the oxen carts. And against these odds, she is fighting, fighting with hand canons, presumably made in her own Ironworks, fighting against nature to carve out a piece of its land for herself and her people.

Miyazaki's characterization of Lady Eboshi, and through her, the community and spirit of Irontown as the embodiment of twentieth-century ambition is present in every action taken by her throughout the movie. Irontown is a utopia, a dream realized by Lady Eboshi, a woman with self-acquired power, using said power throughout the movie to tame her surroundings. With her hubris and her over-confidence in both her ideals and her technology, one cannot help but think of how much humanity will change nature as well as its own nature in the future. Nature is against Lady Eboshi both physically and spiritually. For Lady Eboshi, nature is something to be tamed, claimed, and brought to its feet. But in the end, like everything in this world, such acts of hubris have their price. Lady Eboshi's price was her arm and her pride. Miyazaki seems to ask us: how much is humanity willing to pay to be able to bend nature to its will?

Fear and Fury in the Forest of the Shishigami

Throughout the movie, it is shown that the residents of Irontown fear the forest, despite wishing to tame it. To explain this dichotomy, let us return to a part of the quote in the introduction of this chapter. Here Miyazaki laments to Satō that "I think we [the Japanese] have always been cruel toward nature," going against the commonly held notion that the Japanese as a people have a distinct relationship with nature. Miyazaki's lament is reminiscent of an observation made by anthropologist Arne Kalland regarding nature in Japanese culture:

> The Japanese have, like most people, an ambivalent attitude toward nature, in which their love for nature is only one dimension. But they also fear nature. They have learned to cope with natural disasters caused by earthquakes, erupting volcanoes, typhoons, and floods. But the threat of nature goes beyond this. Many Japanese seem to feel an abhorrence toward "nature in the raw" (*nama no shizen*) [...] and only by idealization "taming" (*naruse*)—e.g., "cooking" through literature and fine arts, for example—does nature become palatable. (Kalland 1995, 221)

This dichotomy of humanity's relationship to nature is what Miyazaki tries to portray throughout the movie, be it in Japan or elsewhere, humans might love nature while at the same time fearing it when it appears in its most sublime and raw form. Lady Eboshi embodies the twentieth-century view of nature as something to be conquered and tamed. When she is about to fire and kill the Shishigami, the embodiment of raw, untamed nature, she exclaims: "Now watch closely, everyone. I'm going to show you how to kill a god. A god of life and death. ..."

Raw nature, untouched and unwilling to be tamed, is always present in *Princess Mononoke*, while simultaneously, its taming and destruction are major points of conflict throughout the plot. The hands of humanity in the form of Irontown have influenced the whole ecosystem of the Forest of the Shishigami. The disturbance reaches as far as Ashitaka's village where the boar *kami* Nago, turned *tatarigami*, with terrible scars and black phantom-like worms protruding out of its body, attacks the Emishi. Raw nature is represented by the different *kami*: the wolves, led by Moro; the boars, led first by Nago, and later by the Okotto; and the elusive *kodama* (tree spirits) that accompany the Shishigami, the deer-like kami, of the forest. This somewhat un-organized group of characters can be named the third community of the movie: the Forest.

The Shishigami, the mysterious, deer-like ruler of the forest is portrayed as an elusive yet powerful being. Its face is eerily reminiscent of an old man, its antlers are like tree branches and its mane is like a long beard. A shapeshifter, during the day it is in its deer-like form, while at night it changes into an enormous, transparent giant, the Daidarabotchi. Throughout the movie, it is shown that the Shishigami has healing powers, and upon every step it takes, flowers bloom and then wither away, foreshadowing its power over life and death.

While Moro, the wolves, and the boars are the most active, anti-human force in the movie, the Shishigami for a long time does nothing, representing nature's apathy towards human affairs and morality. Via apathy, the Shishigami throughout a majority of the movie embodies the sublime, the fearful beauty and majesty of nature with a mind unknown to mere mortals. None of the inhabitants of the forest, neither Moro, Okotto, or the *kodama*, know what the Shishigami is thinking. Why doesn't it act or seem to care that the forest is dying or being attacked? It is not until its head is cut off that it starts to move, rampaging through the forest, killing everything in its wake. The Shishigami when it loses its head loses its natural state of tranquil apathy and becomes a cruel, dark and headless Daidarabotchi that kills indiscriminately, much like a volcano erupting and destroying its surroundings, with no thoughts behind these actions, as if the anger of the Earth itself is enacting punishment upon its surface.

Throughout the movie *kami* who turn into *tatarigami* become dark, and black phantom-like worms crawl out of their body. Likewise, Ashitaka's wound turns black and oozes out these worms. When asked about this choice of visualization, Miyazaki explains: "When I am suddenly struck by a sense of vehement fury, I feel like something black and viscous is oozing out of my pores and other holes of my body" (Miyazaki 2014, 52). The boar Nago turned *tatarigami* at the beginning of the movie, whispers as he dies that "You shall know my hatred and grief." Hatred and grief turn into fury. This fury is another main theme throughout the movie: every character has some fury in them and how they deal with it symbolizes their relationship with nature.

Lady Eboshi's fury is one of ambition, self-determination and social justice for the unrightfully marginalized. Ashitaka's fury is one of grief and sorrow, both the grief for himself, his future death from the *tatarigami's* curse and the sorrow that is part of his village and his people's history. The third fury is the raw and unbridled fury of nature embodied by San, more widely known under the moniker Princess Mononoke (*Mononoke-Hime*, literally "vengeful spirit-princess" in Japanese). *Mononoke* (lit trans: "mysterious things") is in Japanese literature and folk tradition a word used to describe a wide variety of mysterious and harmful occurrences. Later it became the name used for malignant and vengeful spirits with a will to cause strife, disease, and hardship to others (Foster 2015, 14–15). As the forest's vengeful spirit personified, San truly embodies her namesake.

Throughout the movie San is portrayed as a rough and ferocious person. She was as a baby abandoned in the forest to be eaten by wolves. But instead Moro the wolf adopted her, bringing her up as her own daughter. San and her mother are the second and third female leaders in the movie, both fighting against Lady Eboshi and her Irontown. Regarding her characterization, Napier notes that San's femininity is subdued, and the character's design is rough and animalistic, but not wild and monstrous. She is untamed like the nature she inhabits (Napier 2001, 481).

A shallow analysis would have claimed that San is by design the thematic opposite of Lady Eboshi, but this would be false, for Lady Eboshi, despite her visual beauty, is also untamed by civilization. Both San and Lady Eboshi protect the abused and feel a strong sense of justice in response to the destructive ways of human nature, whether social injustice (Lady Eboshi) or human destruction of nature (San) (Vernon 2018, 121). However, unlike Lady Eboshi's restrained clothing, San's clothing is loose and suggestive. Napier suggests that San's design invokes intense sexuality; the fur in her coat around her neck suggests genitalia, her movements and actions signifying animalistic activities that have an odor of wild "primitive sexuality". The blood that often is smeared on her face, suggest "menstrual blood and aggressive sexuality" (Napier 2001, 481). I find this interpretation interesting, yet I would argue that rather than sexual themes, San's design is Miyazaki's way of visualizing San's fury; they are signifiers of her animalistic bloodthirst, her unbridled anger toward humans, embodying the fury of nature. Anger that consumes her.

The ending of the movie can therefore not be seen as a happy one. As Kristen L. Abby notes, even when San and Ashitaka want to be together (their union would symbolize the ultimate union between untamed nature and tamed humanity), San is unable to forgive humanity: "In *Mononoke-hime* there is no hero to rescue humanity, the natural world is both with and against us, and our best attempts at social justice depend on degradation of another" (Abbey 2015, 118).

Ashitaka's relationship with nature is displayed more clearly in discussion with the other two main characters. Throughout the movie Ashitaka is shown to feel close to nature, but unlike San's wildness, his relationship is one of understanding and cohabitation. The curse put on him was not to the result of him wanting to hurt the boar *tatarigami* but because he wished to protect his village, paying whatever price necessary. Ashitaka was following the logic of nature and is likewise following it to its logical conclusion as he tries to find the source of the curse. When Lady Eboshi moves to kill the Shishigami, Ashitaka tries to persuade her to try and live together with the forest. But in the end, no one listens to him, and their fury overtakes their reason. Ashitaka's character embodies the sympathetic and non-ambitious side of humanity, while the curse mark that scars his body symbolizes humanity's cruelty and desire to bend nature to its will, for its cause is the very ambition that drove Lady Eboshi to shoot the boar god.

Princess Mononoke depicts the dangers of human ambition and how it courses through history like a *tatarigami* cursing everything it touches. The movie shows that the ideas that started the industrial revolution in Europe during the 1700s, starting the greatest project of man's destruction of nature are not a isolated example of historical hubris. Humanity's fear for nature and taming of both nature and this fear in the name of progress have been a leitmotif throughout human history. Miyazaki seeks to tell us that humanity's fear of nature is nothing new and that it is this fear that has brought industry and progress to us, but at the same time it has created a lot of grief and sadness. Here the movie is less historical fiction and more a mirror to history.

One can possibly "listen" to the movie's three conflicting relationships with nature via the lens of Gunlög Fur's concept of *concurrences*. This concept allows us to "listen" to complementary, competing, or conflicting occurrences in history and unearth different voices of both those in power, the furious and the dispossessed, thereby allowing all "sides" to have a voice when these different voices encounter each other (Fur 2017, 39–40). If Lady Eboshi is the voice of the ideal historical twentieth-century person, the ambitious and nature-taming human, whose claim to authority and historicity is a given, and San voices the violent fury of those who do not wish to be tamed; Ashitaka embodies the concurrent voices of the marginalized and dispossessed whose calls for change and empathy have been ignored by those in power.

Concluding Thoughts on Miyazaki's Environmental History

> Since humans are so cruel, I have tended to depict nature in a gentle way, but nature itself can be brutal. It can be irrational. It can be very capricious as to why one organism stays alive and another organism dies. Nature is totally indifferent to the good and evil of individual organisms—Miyazaki Hayao interviewed by Fujiki Kentarō, 1997 (Miyazaki 2014, 39)

When the head of the Shishigami is severed by Lady Eboshi's musket shot, the *kami* turns into a headless Daidarabotchi, a malformed and hideous giant, mindlessly destroying everything in its path, invader and forest life, everything gets destroyed. Similarly to how it ruled the forest indifferently, it now destroys it with the same indifference, endangering both those who wished to kill it and those who wished to save it; all must perish. This change in nature is the culmination of the movie's historical environmentalist philosophy: that humanity has always been cruel and nature indifferent. For, in summary, Miyazaki's environmentalist story telling is one of cruelness, and brutal, unforgiving fury that unrelentingly courses through the veins of history.

Miyazaki's movie goes against the historical assumption often proselytized by *Nihonjinron* and less radical movements, who often seek to promote what Ugo Dessì calls Japan's "eco-nationalism," the appropriation of environmental causes by Japanese institutions to rebrand both historical and current pollution and create an "updated image of themselves" and appeal to a more eco-conscious international audience (Dessì 2017, 92–93). *Princess Mononoke*'s historical framework and its fictional setting are a conscious attempt by Miyazaki and his team to combat and discuss these historiographies head-on.

A major theme and historical lesson found in the movie is that no history is environmentally friendly, no culture, no nation, no country, and no people have lived together with nature in the way romanticized nativist ideologies would have us believe. The Japanese throughout history, like people elsewhere, have polluted and destroyed nature as much as any other major group of people. This incorporates a dimension of Miyazaki's own environmental historiography: that humanity has always been cruel and brutal to nature, and nature has been cruel and brutal back. The mark of the *tatarigami*, the spirit of hatred and grief, vengeance and fury, might never truly fade away: its shadow will haunt humanity till the end of times.

But, like Ashitaka, who leaves everything behind to find answers and gives everything up to help strangers in need; just because things might look hopeless does not mean we should just let things be and accept the status quo. Fear and fury might be two of the most primal emotions humans embody, but Ashitaka shows us throughout the movie that these emotions do not define us, as he does not let his cursed mark define his actions. Despite the cruelty and brutality of our own nature and the nature around us. Despite the indifference of nature's spirit. The Shishigami makes flowers bloom.

Notes

1. This conversation, first published in a special edition of Kinema Junpōsha entitled "Miyazaki Hayao to Mononoke Hime to Studio Ghibli" ("Miyazaki Hayao and Princess Mononoke and Studio Ghibli") is translated into English and available in a collection of essays, interviews, and articles by or with Miyazaki Hayao in Miyazaki 2014, 42–59. Throughout this paper, most dialogues and interviews with Miyazaki Hayao are quoted from this book. The talk is in this translation titled: "You Cannot Depict the Wild Without Showing Its Brutality and Cruelty: A Dialogue with Tadao Satō." This paper's title is inspired by this title.
2. Research on Miyazaki Hayao's work exists in abundance, to the extent that scholars have lamented that the works by Miyazaki overshadow scholarship on other Japanese animation (see Thomas 2012; Berndt 2018). Beyond the

books and articles used in this paper, let me highlight some studies that might be helpful for further research: for an overview of the history of anime, with a chapter especially on *Princess Mononoke* see Napier 2005. For a collection of current scholarship, see Denison 2018. For a comprehensive study on the overall audience reception of the religious elements in Studio Ghibli movies, amongst these *Princess Mononoke*, see Ogihara-Schuck 2014.

3. *Kami* is an elusive concept best not translated as it then may gain connotations that it does not historically carry. For a short summary of what might fall under this concept, see Brian Bocking (1995, 84): "Kami may refer to the divine, sacred, spiritual and numinous quality or energy of places and things, deities of imperial and local mythology, spirits of nature and place, divinized heroes, ancestors, rulers and statesman."
4. The name "Shishigami" is curious. The silhouette of the character is reminiscent of a deer (*shika* in Japanese) and in the English translation its name is translated into "deer god/spirit," but its name is not "Shikagami" in Japanese. "Shishi" means lion but can denote the ruler of the whole animal kingdom in general in older contexts. According to Ōta Tetsuo, an explanation for this word choice is that the word "shishi" in older times meant beasts of the forest in general, thus counting deer, boars, and other beasts under the same word. Therefore, a more apt translation might have been "beast god/spirit". For a more throughout speculation on the "beast" naming choices made by Miyazaki, see Ōta 2005. Another possible connection can be to the *kagura* (a dance performed at shrines and festivals) *Shishi-odori*, the "deer dance" from Iwate prefecture, where the masks the performers wear look less like deer heads and more like the traditional *shishi-mai* ("lion dance") mask.
5. Interestingly the text in the intertitle of the movie is very folkloresque and ahistorical in its construction. The original Japanese intro starts with the following phrase: "Mukashi, kono kuni wa fukai mori ni ōware, soko ni wa taiko kara no kamigami ga sunde ita" which can be translated as: "In olden days, this country was covered with deep forests where the *kami* of ancient times lived." Thus, setting the movie in the frame of a legendary tale and not a historical account.
6. In the original Japanese version, his official name in the movie is *Ashitaka-hiko*, "hiko" in this context meaning "next leader," translated as "prince" in the English translation. As John Tucker suggests, Ashitaka's character and journey are comparable contrast to the legend of Yamato no Mikoto, a pseudo-historical figure in the Yamato hegemony's historiography, turning Ashitaka into an anti-Yamato, for a more throughout analysis of Miyazaki's possible intentions with this comparison, see Tucker 2003.

7. Her name *Gozen* is most likely a double reference, both to famous *onna-musha* ("female warrior") such as Tomoe Gozen (c. 1157–1247) and Fujishiro Gozen (sixteenth-century), thus adding another layer to her character as a warrior; and to famous noble ladies and courtesans who also bore this name, such as Tokiwa Gozen (1123–c. 1180) and Shizuka Gozen (1165–1211), adding to her beauty and connection to warlords.

References

Abbey, Kristen L. 2015. "See with Eyes Unclouded": Mononoke-hime as the Tragedy of Modernity. *Resilience: A Journal of the Environmental Humanities* 2 (3): 113–119.

Ackroyd, Joyce. 1959. Women in Feudal Japan. *Transactions of the Asiatic Society of Japan* 7 (3): 43–44.

Berndt, Jaqueline. 2018. Anime in Academia: Representative Object, Media Form, and Japanese Studies. *Arts* 7 (56): 56.

Bocking, Brian. 1995. *A Popular Dictionary of Shinto*. Surrey: Curzon Press.

Denison, Rayna, ed. 2018. *Princess Mononoke: Understanding Studio Ghibli's Monster Princess*. London: Bloomsbury.

Dessì, Ugo. 2017. *The Global Repositioning of Japanese Religions: An Integrated Approach*. London and New York: Routledge.

Foster, Michael Dylan. 2015. *The Book of Yōkai: Mysterious Creatures of Japanese Folklore*. Oakland, CA: University of California Press.

Friday, Karl F. 1997. Pushing beyond the Pale: The Yamato Conquest of the Emishi and Northern Japan. *The Journal of Japanese Studies* 23 (1): 1–24.

Fur, Gunlög. 2017. Concurrences as a Methodology for Discerning Concurrent Histories. In *Concurrent Imaginaries, Postcolonial Worlds: Toward Revised Histories*, ed. Diana Brydon, Peter Forsgren, and Gunlög Fur, 35–57. Leiden and Boston: Brill.

Goto, Michiko. 2006. The Lives and Roles of Women of Various Classes in the IE of Late Medieval Japan. *International Journal of Asian Studies* 3 (2): 183–210.

Hall, John Whitney, and Toyada Takeshi, eds. 1977. *Japan in the Muromachi Age*. Berkeley: University of California Press.

Hudson, Mark J., Bill Finlayson, and Graeme Warren. 2017. Okhotsk and Sushen: History and Diversity in Iron Age Maritime Hunter–Gatherers of Northern Japan. In *The Diversity of Hunter Gatherer Pasts*, ed. Bill Finlayson and Graeme Warren, 68–78. Oxford: Oxbow Books.

Inoue, Tatsuo. 2010. Tatara and the Japanese Sword: The Science and Technology. *Acta Mechanica* 214 (1): 17–30.

Kalland, Arne. 1995. Culture in Japanese Nature. In *Asian Perceptions of Nature*, ed. Ole Bruun and Arne Kalland, 243–257. Nordic Proceedings in Asian Studies 3.

Kenichi, Iida. 1980. *Origin and Development of Iron and Steel Technology in Japan.* Experience of the UNU Human and Social Development Programme Series 8.

Kozuka, Jukichi. 1968. "Tatara" Process: A Pig Iron-and Steel-Making Process, Transmitted from Ancient Times in Japan. *Transactions of the Iron and Steel Institute of Japan* 8 (1): 36–47.

Kurushima, Noriko. 2004. Marriage and Female Inheritance in Medieval Japan. *International Journal of Asian Studies* 1 (2): 223–245.

Matsumura, Hirofumi, and Yukio Dodo. 2009. Dental Characteristics of Tohoku Residents in Japan: Implications for Biological Affinity with Ancient Emishi. *Anthropological Science* 117 (2): 95–105.

Miyazaki, Hayao. 2014. *Turning Point 1997–2008.* Translated by Beth Cary and Frederik L. Schodt. San Fransisco: VIZ Media.

Napier, Susan J. 2001. Confronting Master Narratives: History as Vision in Miyazaki Hayao's Cinema of De-Assurance. *Positions: East Asia Cultures Critique* 9 (2): 467–493.

———. 2005. *Anime, from Akira to Howl's Moving Castle. Experiencing Contemporary Japanese Animation.* New York: Palgrave.

Niskanen, Eija. 2019. Deer Gods, Nativism and History: Mythical and Archeological Layers in *Princess Mononoke.* In *Princess Mononoke: Understanding Studio Ghibli's Monster Princess*, ed. Rayna Denison, 41–56. London: Bloomsbury.

Ogihara-Schuck, Eriko. 2014. *Miyazaki's Animism Abroad: The Reception of Japanese Religious Themes by American and German Audiences.* Jefferson: McFarland and Company.

Ōta, Tetsuo. 2005. On *Princess Mononoke. Obirin Review of International Studies* 17: 8–21.

Thevenin, Benjamin. 2013. *Princess Mononoke* and Beyond: New Nature Narratives for Children. *Intersections: Studies in Communication and Culture* 4 (2): 147–170.

Thomas, Jolyon Baraka. 2012. *Drawing on Tradition: Manga, Anime, and Religion in Contemporary Japan.* Honolulu: University of Hawai'i Press.

Tonomura, Hitomi. 1990. Women and Inheritance in Japan's Early Warrior Society. *Comparative Studies in Society and History* 32 (3): 592–623.

Tucker, John A. 2003. Anime and Historical Inversion in Miyazaki Hayao's Princess Mononoke. *Japan Studies Review* 7 (1): 65–102.

Vernon, Alice. 2018. Beyond Girlhood in Ghibli: Mapping Heroine Development against the Adult Anti-hero in *Princess Mononoke.* In *Princess Mononoke: Understanding Studio Ghibli's Monster Princess*, ed. Rayna Denison, 115–130. London: Bloomsbury.

BY

PART IV

History, Speculative Fiction and Real-World Social Change

CHAPTER 13

Intervening in the Present Through Fictions of the Future

Kristín Loftsdóttir

Introduction

A small robotic creature piles up endless stacks of trash in a desolated world. It is deprived of anything except fields of debris and trash that seem to stretch on forever. Huge dysfunctional TV screens show glimpses from the past—the screens being themselves left-overs of a life that once was—that presumably is our future, explaining the impending evacuation of the earth due to a manmade environmental catastrophe, where humanity will wait in space for the earth to heal and for it to be inhabitable again. The evacuation has clearly happened already, with no humans having returned for some reason. In a sense, all humans have been rendered refugees—and while the movie implies that this is due to generalized human actions and overconsumption, recent critical perspectives on the concept of Anthropocene reflect both the racialized and deeply unequal burden of the climate crisis, as well as responsibility for how it came about (Park and Greenberg 2020). Regardless, the first scenes of the film *WALL-E* (Stanton 2008)—premiering in 2008—are characterized by a deep sense

K. Loftsdóttir (✉)
University of Iceland, Reykjavík, Iceland
e-mail: kristinl@hi.is

J. L. Hennessey (ed.), *History and Speculative Fiction*,
https://doi.org/10.1007/978-3-031-42235-5_13

of loss. It is a sense of loss for something that has not yet taken place but seems inevitable. We have seen this narrative repeated in different forms; in fact the disaster where human beings end all inhabitation of the earth has been shown in so many forms that it almost seems to have happened already. The sense of the future as a catastrophe that is unavoidable is clearly seen in scholarly theories that have difficulties seeing solutions to our impending problems, especially in regard to the climate crisis. Sociologist John Urry remarks that there are "no good outcomes, only degrees of bad" (quoted in Tutton 2017, 489). In both a humble and powerful way, *WALL-E* reflects the intersecting crisis and "crisis-talk" characterizing the twenty-first century (Loftsdóttir et al. 2018), where the future is uncertain and even already lost. The phrase "crisis-talk" refers to engagements with diverse threats to the future—real or not—whether different types of environmental disasters, crises of multiculturalism, or economic stagnations, often due to a fight over scant resources. Precarious migrants—including refugees, asylum seekers, and economic migrants—are also often presented as a direct threat to the Global North's future.

This chapter focuses on fictions about the future as one site of intervention where key questions are voiced regarding social justice and who is entitled to what rights. Fiction about the future does not necessarily involve a vision or even a prediction of the future but can be seen as using the future as a tool to think about the present. Some of current fiction's depictions of the future and its intersecting crisis thus seems to revolve around a critical intervention into "present-day realities" (see Chaudhuri 2011, 191), making the presence visible in a new and sharper light. My discussion uses the concept of *concurrences* to think about narratives about the future in popular fiction, where concurrences can be seen as inviting an analysis of narratives or histories as concurrent that are usually seen as separate (Brydon et al. 2017, 3). The concept is multilayered, referring not only to a theory but a methodology for "identifying connections across categories that once existed as separate areas of investigation that seemed capable to sustaining inquiry wholly within themselves" (Brydon et al. 2017, 6). As Richard Tutton points out, the future can be seen as simultaneously "material and discursive," meaning that it should not be conceived of as a representation but as "enacted in practice" (2017, 485). Through the enactment of the future in various popular fiction, I see the future as acted on and intervened with. Using concurrences—here referring to the placement of different narratives of the future next to events taking place in the present—can be seen as a part of the enactment of the future in the present.

The future itself has long been a tricky subject for scholars to analyze conceptually and empirically as reflected in the work of Richard Tutton (2017), who uses the phrase "wicked futures" to capture these difficulties for scholars in analyzing the future. The future is always something that has not happened, meaning that it is "wicked" in the sense that it is obscure and "difficult to do something with" (Tutton 2017, 480). I stress here how many fictions about the future are not necessarily about the future but rather, in the words of Ziauddin Sardar, time and space in the works are often "window dressing" (2002, 1). The placement of the future and present side by side creates a space of reflection and critical interventions on the present and the actual predicted future. To quote Sardar's work again, in the case of science fiction, even though appearing at first glance to be concerned with space and faraway galaxies, in fact "the space that science fiction most intimately explores is interior and human" (2002, 1).

My analysis of narratives about the future addresses different materials, both textual and visual. I do not try to analyze these materials in their complexities or totalities, but rather to give personal insights into how they *can* be interpreted as an intervention into present discourses. Furthermore, creative fictions—textual and visual—are also objects that travel and are mobilized at different times, and I will not limit my discussion only to narratives written in the present for the present, but also give some reflections on narratives coming from the past, which can be used to think about particular aspects in the present. Thus, the discussion critically asks how examples from fiction about the future can help us to gain deeper insights into some of the key issues that scholars deal with today, such as the "categorical fetishism" (Crawley and Skleparis 2018) often strongly characterizing discussions of refugees and asylum seekers.

I start the discussion with focusing on some of the contemporary proclamation of crisis, stressing in particular the so called "refugee crisis," where there was both a portrayal of refugees as posing a risk to the future, and a reaction to their dehumanization and the security measures taken against them. Then, I ask how current science fiction can be seen as intervening in discussions of refugees and crisis, and thus conceptualized as concurrent with these, and thus as more concerned with the present than the future. Finally, I draw attention to the role of the android as a figure that is good to think with, in regard to what it means to be human and to intervene in issues of the present.

The Future and Crisis

Walt Disney's theme park Disneyland is probably one of the clearest examples of how the idea of modernity stood for a promise of an almost magical future where technology was the key to prosperity, freedom, and the well-being of all (Loftsdóttir 2021). As reflected in that parts of the park were even named "Tomorrowland," it provided an optimistic view of the wonders waiting in the future. The collapse of this sense of the future is clearly signified in the artist Banksy's inversion of Disneyland, the theme park Dismaland where we see contemporary horrors displayed in different theme park scenarios presented by different artists. At the park, visitors experience securitization in action, as well as seeing displays with little boats crammed with refugees, a large distorted mermaid, as well as reminders of the horrors of meat industries in butchered horses at the carousels. Dismaland's overall atmosphere is of dark colors and shabby surroundings, with sullen, non-smiling staff, capturing the dream of modernity as worn out and/or transformed into dystopia (Loftsdóttir 2021). As such, it speaks to a decade that has been referred to as our "dark times" (Cantero 2017), reflected in the multiple eroding of rights through the process of neoliberalism, involving, among other things, a loss of welfare benefits and pension rights. These have gone hand in hand with an extreme concentration of wealth in the hands of few, along with a loss of faith in democracy itself (Moore 2018).

I have used the concept of "cancellation" to capture the essence of diverse and intersecting crises where it is impossible to predict when things will again go back to "normal" or if the cancellation will continue indefinitely. The concept of cancellation reflects how aspects that were associated with the future for many in the Global North, such as stable jobs, home-ownership, and consumption of various kinds, are now perceived as being at risk (Loftsdóttir 2019). The disappearance of predominant visions of this anticipated future of modernity has, as Andrea Muehlebach (2013) has argued, created a sense of loss for those who celebrated the project and process of modernization but also those who criticized its premises and content (see also Bauman and Bordoni 2014).

When the term "crisis" is evoked, it is important to engage with it analytically, asking critically why it is evoked and what proclaiming "a crisis" does (Loftsdóttir 2016). A part of the allure and risk of use of the term crisis is that as a term it often seems self-explanatory (Roitman 2013, 3). Drawing on Mary Douglas' words in regard to "risk," the term "crisis" can be appealing since it is an abstract concept, universalizing, and powerful in its succinctness (Douglas 2003: 15). The so-called migration crisis clearly reflects the need to critically ask why a particular crisis is evoked. Narratives of a "migration crisis" have been visible in the European context for quite some time—earlier revolving around the crisis of multiculturalism (see the criticism of Lentin and Titley 2012)—but claims of a "refugee crisis" became particularly salient in the early and mid-2010s. In the aftermath of the so-called "Arab spring," people were fled from Libya and Syria as a result of a civil war that started in 2011. In Libya, these were not only civilians escaping but also citizens from other African countries staying in Libya for work (Morone 2017). The Italian government declared a state of emergency in 2011, with the E.U. reacting to this humanitarian crisis by sealing its borders. Scholars have pointed out that the proclamation of crisis in fact facilitated various reforms, involving different types of border controls and the detention of people seeking international protection (Pinelli 2018; Majcher et al. 2020). Italy and Greece become the main entry points for people fleeing war or escaping other intolerable conditions in 2013 (Pinelli 2018). When in 2015, migrants seeking to enter Europe became increasingly visible, the rhetoric of Europe as under "siege" due to illegal or criminal populations intensified as well (Hage 2016: 39).

The so-called refugee crisis in Europe in the mid-2010s also made more visible to many living in Europe the growth of the security state. This has involved the cancellation of civil rights and liberties of racialized others under the pretext that this was necessary to protect the citizens of the states in the Global North. Scholars have pointed to the "War on Terror" in the U.S. as important in enabling authorities to act on suspicions against potential enemies and invalidate the basic rights of potential suspects. Subjects were regularly defined as "unlawful combatants" or "detainees," which excluded them from the protection of the Geneva Convention for prisoners (Chaudhuri 2011, 193). Through the proclamation of State of Exception, return to systematic violence and torture by the U.S. state was justified (Puar 2018, 113; Höglund 2017, 290–291). Within Europe, scholars and activists criticized how the strengthening of institutions such as Frontex has led to a militarization of Europe's borders (Pinelli 2018, 729), where the distinction between rescue operations and the pushback of migration becomes blurred (Davitti 2019, 1175). "Crisis talk" was important in justifying such pushback actions and securitization, as crisis talk in general stimulates and calls for affective reactions, reactions that can mobilize support for various state policies and interventions. As phrased by Loftsdóttir et al.: "[C]risis talk can be one approach to win social consent and build or reactivate a certain common sense" (Loftsdóttir et al. 2018, 22).

The proclamation of a refugee-crisis in the aftermath of the economic crisis of 2008 facilitated, furthermore, the increased infiltration by populist groups of the political sphere in Europe (Decker 2016) by making it easier to imagine the nation as under threat by racialized others, i.e., refugees (Thorleifsson 2018). Within different discourses, "economic anxieties" were translated into "ethno-religious grievances" (Thorleifsson 2018). As Hakki Taş (2020) shows, populist leaders seek to regulate time in particular ways where different segments of time are stretched, compressed or hidden from view. Populists' predictions of the future can thus be seen as generating a vision of an "alternative world" of the future—i.e., a world of chaos, where the racialized others have taken over (Loftsdóttir 2019). This sense of crisis has contributed to increased polarization in the social sphere where refugees are framed either as a threat or as victims (Hameleers 2019, 219).

As indicated earlier, expressions of solidarity with refugees were also quite significant, sometimes along with strong criticism of the European refugee system and securitization. Particular events such as the death of the toddler Aylan Kurdi in 2015 have been seen as causing a paradigm shift in general discussions about refugees (Siapera 2019, 248), reflecting how solidarity with refugees has been shifting in accordance with particular events popularized through the media (Brändle et al. 2019, 722). Nevertheless, in the mid-2010s, heated debates have taken place around the issues of solidarity with refugees, actions on Europe's external borders, and the responsibility of different member states by politicians and the public alike (Brändle et al. 2019; Bock and Macdonald 2019).

Thinking About the Present Through Science Fiction

So how can we think through and across the "crisis" of migration by using fiction about the future? The concept of ***concurrences***, first of all, draws attention to the fact that some of these fictions—while taking place in the future—are convergent and entangled with the present. Gunlög Fur's (2014) work shows how the tendency to treat particular histories as separate—even though they actually take place in the same time and space—often conceals entanglements and power imbalances. The scholarly project of concurrences must thus partly involve exposing these power imbalances. Similarly, fiction about the future that tells stories to be concurrent with particular events in the author's present can be seen as taking part in exposing particular power dynamics and inequalities. Furthermore, by locating their narratives as taking place in the future, authors manage to move beyond dull party-political debates into the core of the issues at stake. The future or alien setting can thus more be imagined as a backdrop—or "window dressing" in Sardar's (2002) words—making it easier in some sense to insert charged political issues into popular discourse.

I want to start my discussion with Shohini Chaudhuri's (2011) excellent analysis of the film *Children of Men* (Cuarón 2007), where the intention is clearly to draw attention to contemporaneous hostile migration regimes or to the context of when the film was made, i.e., "The War on Terror." The film's central plot revolves around the apocalyptic vision of a near future where sudden and unexplained infertility strikes the world. It simultaneously locates the story within extremely hostile U.K. government actively criminalizing and tracking down migrants. This position of migrants is not so much explicitly explained in the film—and in fact, the main protagonist, Theo, seems to avoid looking at what is happening around him—but rather the migrant part of the story takes place in the background. Nevertheless, Slavoj Žižek has explained that the background of the film is actually its "true focus" (see discussion in Chaudhuri 2011, 191). Thus, alongside scenes somewhat typical of action films where Theo is trying to save a woman who is the first human to become pregnant in years, the background depicts chilling but familiar images of racism, such as migrants in cages awaiting deportation and references to photographs taken in Abu Ghraib prison, including the "hooded man" (see Chaudhuri 2011, 199). Through these background images, as well as various powerful metaphors within the film itself—such as that the woman who becomes miraculously pregnant and thus the last hope to save humankind is a Black refugee—the film can be seen as using concurrences as a method in various senses: by intervening in its current political context, and by placing two stories (the one in the front and the one in the background) side by side. Scenes from the recent present known to the viewer, such as references to torture, are inserted clearly in the narrative.

The movie *Children of Men* vividly shows how the cancellation of the human rights of some people through extreme measures can appear acceptable to some because they believe that they will not be affected themselves. Or as phrased by Chaudhuri, it involves "activating traditional forms of racism in the belief that only 'others'—the Muslims, Arabs, Asians and Blacks—will be affected" (Chaudhuri 2011, 194). The film clearly demonstrates the shared vulnerability of both those defined as others and those who are not (Chaudhuri 2011, 201). In this film the populist vision of the future can be seen as turned upside down with the security state and tough measures on migration, creating an alternative world (Loftsdóttir 2019), but exposing the links to totalitarian regimes and the vulnerability of citizens and migrants alike. While I position this film mainly as intervening in its present, narratives of the future of course do not only cross space but also time. The theme of *Children of Men* became even more relevant ten years later, or as one film critique proclaimed in the mid of the migrant crisis during the year 2016, the film's "version of the future is now disturbingly familiar" (Barber 2016).

The film *Valerian* (Besson n.d.), which also came out in 2017, is made within a completely different genre, characterized by a fast and action-packed story for a younger audience. The film is full of aliens and androids and contains no complex discussion of their boundaries, nor humanity's boundaries with alien others. The story centers to some extent around the love relationship between the two main characters Valerian and Laureline but is mainly an action film. While the comic book series that the film seeks inspiration from, published in the 1970s, predicts an apocalyptic future for humanity, we see here a distant future of friendly relations with different species—humanoids and others.

Most of the film follows a somewhat standard storyline for such action films with images of dance clubs, a car chase, and so on. While not strongly emphasized for most of the film, a story about justice and refugees can still be seen as embedded in the plot. The planet Mül was destroyed several decades earlier as unnecessary collateral damage in a war between humans and alien powers.[1] At the end of the film, the human officer who ordered the destruction of the planet Mül explains that he needed to kill all of the few survivors from the destroyed planet to hide that that this total destruction was not necessary, but rather that the human leaders decided to sacrifice Mül's inhabitants as it was more convenient way to win the war. This has to be hidden in order for his people to avoid being liable to pay reparations. Significantly, humanity's supremacy in galactic relationships would also be weakened if the truth came out. He explains that: "Our council saw fit to protect our citizens first and foremost..." He asks Valerian and Laureline: "would you [...] risk wrecking our economy for the sake of a bunch of...." When his voice trails off Laureline adds questioningly and provocatory: "savages?" Here it is tempting to draw a parallel between the economic arguments of populist leaders at the time in justifying the inhumane treatment of refugees and their detention and exclusion from the space of Europe under the slogan "our people first" (Hameleers 2019, 813). Part of the film's message at the end is an emphasis on "doing the right thing" where sometimes even the laws that people respect need to be broken to do what is right. The film's ending reflects this when Valerian and Laureline break the law of their government to help the refugees to have a future, and thus to follow what they know is the right action when the law fails them and the survivors from Mül. It should be pointed out that even though making this point, the film can also be criticized for various stereotyping such as the simplistic portrayal of rigid ethnic boundaries, the use of different racist Western imaginary of African savages, where some are noble and other ignorant and laughable.

The political environment in which the film was made was not simply characterized by hateful discussions about refugees by populist leaders, but also strong criticisms of Europe's regime of mobility, which facilitates the mobility of some, while rendering others immobile, as well as the criminalization of people seeking shelter within Europe. It is difficult not to think of instances where our regime of mobility has sought to penalize people fleeing, but also those who help them. This became particularly evident during the crisis in 2016, where ordinary citizens in Europe were arrested and punished for actions like giving hungry people food and driving old or tired people short distances, which has been referred to as the criminalization of solidarity (Fekete 2018).

Androids and Being Human

Androids are a part of many fictions of the future and—to refer to Sardar's comments in regard to aliens in general—androids can also be useful to "demonstrate what is not human the better to exemplify that which is human" (2002, 6). The android is not only another version of a monster in popular fiction but rather has an in-between or liminal position as both man and machine and neither of these, which allows for complex and creative questions. For example, *Star Trek: The Next Generation* often used the android *Data* as essential figure in pondering difficult key questions of what it means to be human along with ethical dilemmas and responsibilities of human beings. As I show in the examples below, we can say in the spirit of anthropologist Claude Lévi-Strauss when referring to animals and totems (1963, 89) that androids are "good to think with" in regard what it means to be a human being.

Some recent science fiction films have engaged critically with their era of dispassion and inhuman bureaucracies through emphasizing compassion, where androids are important way to enter into critical discussion about the present. The dystopian *Blade Runner 2049* (Villeneuve 2017) depicts a technological future of isolated human beings shaped by environmental catastrophe. Androids, or replicates as they are called in the movie, have taken the role of disposable humans. Through visual imagery characterized by haze and dull colors, as well as the somewhat disturbing monotonous soundscape of the film, we get the sense of a world empty of compassion and life. Here, the androids take the position of the subaltern, and like the migrants in *The Children of Men*, the androids are "unpeople," to use Mark Curtis' phrase; their lives are as worthless and expendable (Chaudhuri 2011, 192). In one scene of the film, we are led to an abandoned casino standing empty in a radioactive area. In this space of past luxury and affluence, we hear Elvis Presley sing, which further intensifies the sense of a future lost—a future that is hauntingly familiar as it is our recent past. Contrary to the modernist dream, it is not the more advanced technology that carries hope for our dystopian future, but the acknowledgment of the humanity of others—in this case shared humanity with the replicates/androids. The film does not mention refugees and asylum seekers, which were quite visible in the media at the time of its making, but it is easy to draw that connection from the critical emphasis of the film on securitization and the devaluation of the life of others. The androids/replicates can be seen as standing in for those seeking refuge and new opportunities in the Global North—in both cases the issue being compassion and recognizing shared humanity of some kind. The focus on the aspects of the film which are relevant to the discussion here is not to trivialize how the film also reproduces salient stereotypes, especially in regard to its projection of women as sexual objects tied to men, in addition to depicting graphic violence against women.

Now I would like to turn from fictions of the future as concurrent interventions into the time when they are made toward emphasizing more how the relevance of their critical perspective on dehumanization can cross both time and space. Thus, fictions about the future can be equally or even more relevant one or more decades later as they often try to struggle with broad key questions of being human. The insights of science fiction pioneer Isaac Asimov can be mentioned in this context. Some of his books think through different paths that human societies could take in the future and what that would mean for the kind of lives lived. In my discussion here, I briefly like to mention Asimov's intervention into what constitutes a rights-bearing person, thus giving insights into what scholars have called "categorical fetishism" (Crawley and Skleparis 2018). This term describes how people are seen almost as a different kind of human being if they are classified as refugee or asylum seeker. These are legal categories that have taken on a life on their own when policymakers highlight the importance of distinguishing between people in "real" need and those who are "frauds" and thus not in real need. Asimov's books are famous for his laws of robotics where the robots have three key laws so integrated into their minds that it is impossible for them to kill human beings.[2] These laws can be seen as one of the key premises in many of his fictional works. In *Robots and Empire* (1985), the last book in the Robots series, a group of robots, however, suddenly can and does kill human beings. As is revealed later in the book, robots cannot be changed and these rules cannot be bent to make it possible for the robots to kill a human being. Rather what has been changed is the definition of what constitutes a human being. Or as one character in the novel explains to another, the androids were changed in such a way that they were "geared to respond to a person as human only if he or she spoke with a Solarian accent." Others who did not fit that criteria of speaking with this particular kind of accent were not considered humans and thus disposable (167, 244). I have no idea what Asimov's intentions were when writing this, but to me it captures how legal definitions that seem on the surface to make all human beings equal can still allow for the reduction of some people to non-status or "bare" life, as Giorgio Agamben would phrase it—lives that are seen as lacking value or not deserving political status (see the discussion in Chaudhuri 2011, 192). Categorizing some people as "bogus refugees" reduces potential suspects to bare life, either in their countries of origin or within camps, and stateless persons seem to slip through the cracks of human rights law, somehow not existing at all.

To take another example, *Do Androids Dream of Electronic Sheep?* (Dick 1968), the book on which the 1982 movie *Blade Runner* (Scott 1984) was based, was published in the context of the Vietnam War. The author, Philip K. Dick, said that during this period he felt as if "we had become as bad as the enemy" (Sammon 2007, 243). The story follows Rick Deckard, a bounty hunter, in a post-apocalyptic world, where almost all animals have died out. His job is to find androids that have escaped to earth from the "outer" colonies on other planets. Elaborate tests are used to find renegade androids that are otherwise impossible to distinguish from humans—even when it means that innocent individuals will be sacrificed as well. The androids may act, speak, and feel as human beings but are in fact not, making it is crucial to expose and exterminate them. Dick was notably not using the androids as a metaphor for people discriminated against but rather to symbolize actual human beings who were "physiologically human but behaving in a non-human way" and thus "cruel," "without-empathy," and "less-than-human entities" (Sammon 2007, 244, 262). According to Dick, the context of the Vietnam War made him feel that the important question was not if it was justified to kill people who were so cruel (like the replicants) but rather the dilemma was: "Could we not become like the androids [inhuman, without sympathy] in our very effort to wipe them out?" (Sammon 2007, 244). One of the book's key points can thus be brought to the present by asking critically what it means to refuse the right to life to people who look and act as humans; what does it do to those who are the "real" humans—how does their humanity become disputable as well? It asks critically what kind of society "we" will be left with if everyone is a potential threat, and a potential target?

A similar point is addressed in an episode of *Star Trek: The Next Generation* (1987–1994) where the legal status of Data—the android—is under dispute. A Starfleet officer from the central command wants to claim Data, take him away from his friends and home in order to dismantle and experiment on him in order to gain knowledge that could possibly be beneficial for Starfleet. On the surface, the issue seems to be whether Data should be seen as a human with the rights that this entails or as someone's property, but the episode extends the issue more broadly with questions regarding people defined as disposable. The linking of Data's positionality with slavery and dehumanization of identifiable groups in history is especially evident when one character, Guinan, points at the economic benefits of placing particular people outside humanity, directly connecting to the history of slavery:

> In the history of many worlds there have always been disposable creatures, they do the dirty work; they do the work that no one else wants to do because it is too difficult or too hazardous, and an army of Datas all disposable; you don't have to think about their welfare; of how they feel, a whole generation of disposable people. (Season 2, episode 9, ca 35 min after the beginning of the show)

It is probably no coincidence that Guinan is played by a Black actor, Whoopi Goldberg, which brings more clearly out its connection to the historical legacy of slavery. Like the book *Do Androids Dream of Electronic Sheep?* the episode draws attention to the positioning of people outside the spectrum of humanity as a result of particular practices of categorization and the wider consequences of creating disposable people without any rights or compassion.

To link with Asimov's discussion earlier, Asimov draws attention to how by a slight shift of hand, issues that were seen as intrinsic and non-negotiable are all of sudden possible, as illustrated in today's reality where we see the sudden positioning of refugees outside of basic human rights and obligations. Universal ethical obligations, such as saving someone from drowning or not torturing people, are suddenly set aside as irrelevant.

Concluding Remarks

While the future can constitute a "wicked" subject for scholars to analyze (Tutton 2017), it is a productive resource for various fictional engagements and interventions. As I stress in this essay, one of the magical properties of the future is precisely that it can be a material to "do something with," to use Tutton's phrase in relation to the difficulties in approaching the future itself analytically (2017, 480). In a sense, in speculative fiction, it is not only androids that are "good to think with" but also the future itself.

My discussion here has pointed out that narratives of the future can, furthermore, be a powerful way to talk about the present, making the future concurrent with events in the present, but also engaging with larger questions of responsibility, sympathy, and discrimination that have relevance regardless of their particular historical time. Some of the speculative fiction that I have addressed took place around the same time as the refugees and asylum seekers became strongly visible in the European context in the early and mid-2010s, with a strong emphasis on securitization and portrayal of Europe as under "siege" (Hage 2016: 39). The concerns of many works of speculative fiction with these larger questions of what it means to be human—often through emphasis of non-humans or androids—travel across time and space, as is reflected in how older science fiction often has relevance for issues debated in the present.

Asimov's works draw attention to how key ideals do not have to be changed in order to make the killing or discrimination of others in accordance to the law or universal treaties seeking to protect the right of people, just the key categories that they are based on. The book *Do Androids Dream of Electronic Sheep* along with the film *The Children of Men* pose important questions about the wider effects of dehumanization, which not only harms the victims that it targets but also the wider society that it is supposed to be necessary to protect. Thus, one of the questions posed by these fictions is what the cancellation of the humanity of those seen as "others" does to the future that we are entering and the kind of humanity which we ourselves embody.

Notes

1. This storyline differs in many important ways from the book in which it is loosely based.
2. The first rule states that a robot cannot harm a human being, or allow a human coming to harm; the second that it should follow orders, except when they are in conflict with the first law; the third stresses self-preservation, except when in conflict with the other two laws (see, for example, Clarke 1993).

References

Barber, Nicholas. 2016. Why *Children of Men* Has Never Been as Shocking as It is Now. *BBC Culture,* 15 December. https://www.bbc.com/culture/article/20161215-why-children-of-men-has-never-been-as-shocking-as-it-is-now.

Bauman, Zygmunt, and Carlo Bordoni. 2014. *State of Crisis.* Cambridge: Polity Press.

Besson, Luc, dir. *Valerian and the City of a Thousand Planets.* EuropaCorp et al.

Bock, Jan-Jonathan, and Sharon Macdonald. 2019. Introduction. In *Refugees Welcome?: Difference and Diversity in a Changing Germany*, ed. Jan-Jonathan Bock and Sharon Macdonald, 1–40. New York: Berghahn Books.

Brändle, V.K., O. Eisele, and H.J. Trenz. 2019. Contesting European solidarity during the "refugee crisis": A Comparative Investigation of Media Claims in Denmark, Germany, Greece and Italy. *Mass Communication and Society* 22 (6): 708–732.

Brydon, Diana, Peter Forsgren, and Gunlög Fur, eds. 2017. *Concurrent Imaginaries, Postcolonial Worlds: Toward Revised Histories.* Leiden: Brill.

Cantero, Lucia E. 2017. Sociocultural Anthropology in 2016: In Dark Times: Hauntologies and other Ghosts of Production. *American Anthropologist* 119 (2): 308–318.

Chaudhuri, Shohini. 2011. Unpeople: Postcolonial Reflections on Terror, Torture and Detention in *Children of Men.* In *Postcolonial Cinema Studies*, ed. Sandra Ponzanesi and Marguerite Waller, 191–204. Abingdon: Routledge.

Clarke, R. 1993. Asimov's Laws of Robotics: Implications for Information Technology-Part I. *Computer* 26 (12): 53–61.

Crawley, Heaven, and Dimitris Skleparis. 2018. Refugees, Migrants, Neither, Both: Categorical Fetishism and the Politics of Bounding in Europe's 'Migration Crisis'. *Journal of Ethnic and Migration Studies* 44 (1): 48–64.

Cuarón, Alfonso, dir. 2007. *Children of Men.* Universal Pictures.

Davitti, Daria. 2019. Biopolitical Borders and the State of Exception in the European Migration 'Crisis'. *The European Journal of International Law* 29 (4): 96–1173.

Decker, Frank. 2016. The "Alternative for Germany": Factors Behind its Emergence and Profile of a New Right-Wing Populist Party. *German Politics and Society* 34 (2): 1–16.

Dick, Philip K. 1968. *Do Androids Dream of Electric Sheep?* New York: Doubleday.

Douglas, Mary. 2003. *Risk and Blame.* Abingdon: Routledge.

Fekete, Liz. 2018. Migrants, Borders and the Criminalisation of Solidarity in the EU. *Race and Class* 59 (4): 65–83.

Fur, Gunlög. 2014. Indians and Immigrants-Entangled Histories. *Journal of American Ethnic History* 33 (3): 55–76.

Hage, Ghassan. 2016. État de Siège: A Dying Domesticating Colonialism? *American Ethnologist* 43 (1): 38–49.
Hameleers, Michael. 2019. Putting Our Own People First: The Content and Effects of Online Right-wing Populist Discourse Surrounding the European Refugee Crisis. *Mass Communication and Society* 22 (6): 804–826.
Höglund, Johan. 2017. Can the Subaltern Speak Under Duress?. Voice, Agency and Corporeal Discipline in Zero Dark Thirty. In *Concurrent Imaginaries, Postcolonial Worlds: Towards Revised Histories*, ed. Diana Brydon, Peter Forsgren, and Gunlög Fur, 281–301. Leiden: Brill.
Lentin, Alana, and Gavin Titley. 2012. The Crisis of 'Multiculturalism' in Europe: Mediated Minarets, Intolerable Subjects. *European Journal of Cultural Studies* 15 (2): 123–138.
Lévi-Strauss, Claude. 1963. *Totemism*. Translated by Rodeny Needham. Boston: Beacon.
Loftsdóttir, Kristín. 2016. Building on Iceland's 'Good Reputation': Icesave, Crisis and Affective National Identities. *Ethnos* 81 (2): 338–363.
———. 2019. Crisis, Migration, and the Cancellation of Anticipated Futures. In *Orientations to the Future*, ed. Rebecca Bryant and Daniel M. Knight, *American Ethnologist* website, March 8. http://americanethnologist.org/features/collections/orientations-to-the-future/crisis-migration-and-the-cancellation-of-anticipated-futures.
———. 2021. *We are all Africans Here: Race, Mobilities, and West Africans in Europe*. New York: Berghahn Books.
Loftsdóttir, Kristín, Andrea L. Smith, and Brigitte Hipfl, eds. 2018. *Messy Europe. Crisis, Race and the Nation–state in a Post-colonial World*. London: Berghahn.
Majcher, I., M. Flynn, and M. Grange. 2020. *Immigration Detention in the European Union*. Cham: Springer.
Moore, Martin. 2018. *Democracy Hacked: Political Turmoil And Information Warfare in the Digital Age*. London: Oneworld.
Morone, Antonio M. 2017. Policies, Practices, and Representations Regarding Sub-Saharan Migrants in Libya: From the Partnership with Italy to the Post-Qadhafi Era. In *EurAfrican Borders and Migration Management*, ed. Paolo Gaibazzi, Alice Bellagamba, and Stephan Dünnwald, 129–155. New York: Palgrave Macmillan.
Muehlebach, Andrea. 2013. On Precariousness and the Ethical Imagination: The Year 2012 in Sociocultural Anthropology. *American Anthropologist* 115 (2): 297–311.
Park, Thomas K., and J.B. Greenberg, eds. 2020. *Terrestrial Transformations: A Political Ecology Approach to Society and Nature*. Lexington Books.
Pinelli, Barbara. 2018. Control and Abandonment: The Power of Surveillance on Refugees in Italy, During and After the Mare Nostrum Operation. *Antipode* 50 (3): 725–747.

Puar, Jasbir K. 2018. *Terrorist Assemblages: Homonationalism in Queer Times*. Durham, London: Duke University Press.

Roitman, Janet. 2013. *Anti-Crisis*. Duke University Press.

Sammon, Paul M. 2007. Out Blade Runners, PKD, and Electric Sheep. In *Blade Runner (Do Androids Dream of Electric Sheep?)*, ed. Philip K. Dick, 243–265. New York: Del Rey Books.chay.

Sardar, Ziauddin. 2002. Introduction. In *Aliens R Us: The Other in Science Fiction Cinema*, ed. Ziauddin Sardar and Sean Cubitt, 1–17. London: Pluto Press.

Scott, Ridley. 1984. *Blade Runner*. Warner Bros.

Siapera, Eugenia. 2019. Refugee Solidarity in Europe: Shifting the Discourse. *European Journal of Cultural Studies* 22 (2): 245–266.

Stanton, Andrew, dir. 2008. *WALL-E*. Pixar.

Star Trek: The Next Generation. 1987–1994. Paramount.

Taş, Hakki. 2020. The Chronopolitics of National Populism. *Identities* 29 (2): 127–145.

Thorleifsson, Cathrine. 2018. *Nationalist Responses to the Crises in Europe: Old and New Hatreds*. London, Abingdon: Routledge.

Tutton, Richard. 2017. Wicked Futures: Meaning, Matter and the Sociology of the Future. *The Sociological Review* 65 (3): 478–492.

Villeneuve, Denis, dir. 2017. *Blade Runner 2049*. Columbia Pictures et al.

CHAPTER 14

Building a Kinship Society (Short Story)

David Belden

2109

"Atsa child, what are you going to do with your life?"

"I don't know." The wind was blowing hard on the rooftops, bending the new saplings halfway over. But down here in the grassy alley Atsa and her grandmother were protected by the high buildings either side.

"Well, what do people tell you you're good at?" Grandmother was sitting on their favorite log, the one with a branch that served as a backrest. Atsa sat cross-legged on the ground, massaging one of the old woman's arthritic feet.

"Nothing! I'm no good at the things the clan needs most. Plants don't grow for me. I'm not a good communicator, or a healer. I'm no mechanic, no building maintenance tech. I don't want to get pregnant and have babies sucking on my tits like Keisha. I'm just weird. I love reading old books in the library, but no one tells me I'm good at it. They just get fed up with me for it."

"Well, believe your grandmother: you're good at it."

"Thank you!" Atsa brushed tears from her eyes.

D. Belden (✉)
San Francisco, CA, USA

J. L. Hennessey (ed.), *History and Speculative Fiction*,
https://doi.org/10.1007/978-3-031-42235-5_14

"You know, child, you do have a good feel for feet, and aches, and bones. You know something else? In the old days there were people called 'historians.'"

"I know the word. It means storyteller, doesn't it?" A wind flurry brought the scent of new mown grasses from the upper rooves, where the rest of Atsa's community were hard at work with scythes and twine, loading hay bales onto the chute system that brought them down to ground level.

"Sort of, but there was much more to it. Historians did research: they read old records and tried to work out what had happened in the past. Sometimes their discoveries contradicted what was thought to be true, and brought new thoughts to the people. Historians read books all day, unless they were writing books, or teaching from books."

"Can I become a historian, grandmother?"

"You know well, child, there's no extra food to keep someone who's not doing productive work. So you'll have to be creative. But I want you to think seriously about doing it. Because you never know what will turn out to be productive after all. And you know what we elders say, that we have to heal the seven generations after us, and the seven generations before us."

"They say that, but it makes no sense to heal our ancestors. They're dead."

"It's spiritual work."

"I can't even imagine trying to heal them. To be chained on slave ships from Africa. To be beaten and worked to death and separated from your family. Or to have your sacred lands stolen from you and most of your people killed or die of unknown diseases. It overwhelms me to think of healing the ancestors who went through that."

"You know, you're a light skinned child, and your nose sticks out big and straight."

Atsa looked at the old woman with puzzlement.

"I'm just saying, you didn't mention your white ancestors, did you? You think they didn't need healing too? It's a big question: is it only the people harmed who need healing? If the people doing the harm heal, do they then stop doing so much harm? Isn't that the single major way to prevent future harm?"

"That's what Mom says about Daddy. How much he needs to heal."

"Yes, child. Sadly so. My son is still mad at me and Grandfather. But it's different when people are as angry as he is and yet they belong to a dominant culture that tells them that pointing the finger at the other person is normal. Because then, how would they know they need to heal from blaming and shaming? You know that song, 'When I point my finger at my neighbor?'"

"… there are three more pointing back at me," Atsa did the hand gesture. "Yes, I know. But if you turn your hand knuckles up, you can't see the three pointing back at you. That's being oblivious."

"You've learned your lesson well. Now, it's time for you to go and play, or help with the haymaking."

"I'll go back to the library."

2142

"Atsa, you can't be serious!" Maria protested. They put down the pruning shears they were using to cut back the hedge.

"I'm dead serious. The Begay folk tell me time travel has become a real thing."

"And you want to go back to heal your white ancestors? That's insane!"

"It probably is. But it's also real." Atsa pulled nervously on her braids, not caring how it betrayed her anxiety.

"It's dangerous! How will you get back to us?"

"I might not be able to come back."

"By all that's holy! And you're just telling me now?" Maria stared at her as if she had no words.

"I'm sorry. Grandmother came to me in a dream. She told me to go. I have to."

"This is why you've been practicing gendered language, isn't it. All this 'he / she' shit. I thought you were just getting into the old times for your work. But you've been training to go back in time for months, without telling me."

"I'm scared to death, Maria. Until yesterday it was just a fever dream. I never thought it would be real".

"So why is it real now?"

"Because the elders have chosen me. They say all the research I've done about my ancestors, and that crazy story two of them told, a hundred and twenty years ago, you know what I'm talking about it." She couldn't say it aloud, not at this moment. Her ancestors, Morris Dale and Atsa Begay, had told the world that their lives were altered by a vision of Morris's many-times granddaughter who appeared to them and called them to a different kind of work. In truth, they didn't say a vision, they said a visit. But the historians naturally described it as a vision. And now the elders were telling Atsa she could, in fact, go visit them. "You know, the vision Atsa Begay had, the Atsa I'm named after, the famous one. The elders think it will work. That maybe my visit is why he changed his life. If I don't go, history might have been quite different. We might none of us be here. They might have continued fighting each other, blaming and shaming, until it was too late. And all their signs say I have to leave tomorrow. I'm having panic attacks. I can hardly breathe when I think about the task they've given me. How can I possibly succeed?"

"Don't you know I love you? We're partnered, you can't do this to me!" Maria threw their favorite pot at the wall and slammed the door on their way out. Atsa put her head between her knees and breathed fifteen times slowly.

2022

"Guilty! Guilty! Guilty! We are facing the destruction of our civilization! And these morons, these addicts of God, guns, and opioids couldn't care less! They're as high on their meds as crack addicts on crack. And we know how they despise crack addicts. They lock up Black and Brown people by the tens of thousands. But they're the ones who should be locked up!"

I paused as the crowd yelled their anger. Hard to read those faces. Were they truly angry or just playing up to the spectacle of an old guy ranting?

The grey San Francisco fog streamed over us, looming like the end times themselves. "These morons…" I still couldn't be heard.

My eye was drawn back to an anxious young woman in the second row whose face I had passed over with a jolt earlier, as if, as if…. The shouting died enough for me to yell, "These morons are the ones who are tipping us over the climate cliff, driving our species to extinction! And countless other species!"

"Yeah! Yeah! Preach!" Fists pummeled the air.

How was a 70-year-old white man getting such a rapturous response? An old man with a backache so extreme it was only the pills and adrenalin that were keeping me upright. There were older white folk in the crowd, for sure, but it looked like the young of all hues outnumbered them here. I guess scared young idealists love crotchety, old, principled lefties. I was shamelessly using the Bernie playbook. I'd been called "the Pied Piper of the Angry Ancient Bay Area Left" on one kid's viral tweet.

Another speaker took over. I retreated beyond the crowd to lie on my back on a bench. Connie and Doug from our Grey Disrupters cell had brought a small cooler with ice packs for my back. Bless them. I was staring up at the blowing fog wondering how long I could keep all this ranting up, when a face bent down over mine and blocked the light. Oh God. Her face was a silhouette but I knew. It was the young woman from the second row.

"I can help you," she said. "Just relax."

She moved her hands towards me, and I blurted out, "Don't touch me! I don't like to be touched!"

"I can tell, Uncle Morris. Don't fret. I won't touch you." I raised my head angrily to say, "What's with the Uncle shit?" when I saw the ethereal look that had come over her face. She was kneeling beside me, her hands motionless about six inches above my heart, palms down. Then she began to move them above my body like a blind person feeling a face. Her eyes were closed. Whatever her anxiety had been, she had elevated herself into another realm.

Oh God. A woo-woo freak. Just like her mother. If indeed she was Deliah's. Feel-the-energies-Deliah. Deliah the love of my life. Deliah the woman I had grievously wronged. This one had to be her daughter.

And maybe, my daughter. When Deliah disappeared, I thought she was pregnant. But as hard as I tried to trace her, I never heard a whisper. And now this woman. Her skin was about the right shade, halfway between Deliah's dark walnut and my pasty Anglo. Her nose was surely mine. Up close, she wasn't so young after all, already in her forties maybe. That would be right.

She opened her eyes and rose as easily to her feet as if she was a dancer or martial artist—like Deliah. I whispered, "Who are you?"

"Now you can get up," she replied. "Go on."

I started to move, just to prove how excruciating it would be. But it wasn't. The pain was manageable. I swung my legs to the ground and stood up. I looked down at her: my nose, but Deliah's eyebrows and mouth, the same distance in the eyes, as if both women saw further than the rest of us.

"I knew your mother," I said.

"You have never met my mother. But I know about you and Deliah. Can we go someplace and talk?"

"I have to plan stuff with Connie and Doug. Come with us, explain yourself."

The funky café on Market Street I'd loved was long gone. We ended up in a high-end espresso spot for techies, full of laptops and concentrated solitudes. Many people wore masks, intensifying the separations.

As soon as we sat, lattes in hand, she blurted out, "I'm just going to have to say this straight up. I know you once wrote science fiction. So maybe you can hear me when I say," she took a deep breath, "I'm Deliah's great-great-great-great-granddaughter. Four greats."

"Very funny, Ms. Time Traveler. Who are you really?"

"I'm a victim of hundreds of years of men chasing false gods, and in my time we are paying the price a millionfold."

Deliah wouldn't have been that grandiose, but the daughter, if that's who she was, soon showed she was just as good a storyteller. Not about herself, but about the world. We listened intently as she laid out the history of the coming century: accelerated climate change, the coastal cities half underwater, unstoppable waves of refugees, collapsed economies, warlords, starvation, disease. I could have written it myself.

But then she told another side to the story: the many ways unlikely actors had worked together to curb fossil fuels and create alternatives. In the process they had birthed a new culture of healing for humanity and the biosphere. "We call it the kinship society," she explained. "We have a clan system, and no one is left out of the family. That's why no one is homeless on our streets. The animals, the plants, the land are all part of it. We have powerful laws to prevent people amassing power and wealth, but they are hardly needed now that so many people get the concept of family and kin. People like my father may try to become powerful, but he can't find enough followers. It's the followers who give power to the power hungry."

So she was a utopian activist with a fantasy problem. A joker or mentally ill.

"In our history books, you played a role in creating the kinship culture. If you and others hadn't done that, we would have descended into civil war and horror." She began to weep.

"I don't have time for this. You haven't even explained about Deliah. We lost touch over forty years ago. Where is she?"

Her tears ebbed. "I'm doing a lousy job of this," she gasped. "I need to calm down." She took deep breaths and closed her eyes like she was meditating for a moment, her tears drying on her cheeks. Her face calmed to that woo-woo place I'd seen in her before. She finally asked, "You know what the biggest goal of activism needs to be?"

"What, you're going to give me a lesson in politics?"

"The hardest thing is to fight our opponents and pressure them to do what is needed—like getting America off fossil fuels—while refraining from judging them, while appreciating their wounds, while empathizing with their angers, while treating them as kin. We have to invite them in that loving spirit to co-create solutions with us. Like Gandhi said to General Smuts when he went up against his government in South Africa, "I will win." And when Smuts asked how, and Gandhi replied, "with your help." Which is what happened. In your time it was all laid out and demonstrated by Loretta Ross, who we call a grandmother of the new culture. The culture that brought America and the world together. Do you know her?"

"A nice fairy tale. Try doing that with MAGA Republicans."

"That's what I'm talking about. I believe you can do it, Uncle Morris. At least you can stop abusing them. If you do, you may at least appeal to their more moderate cousins who find you obnoxious."

"Are you kidding me?"

She closed her eyes again. "You're a skeptic, aren't you?" She suddenly appeared formidable, less fantasist than inquisitor or professor. "Not about climate change. But about everything. That's what it was called. A skeptic was someone who had to question everything."

"Yes, I'm proud of that."

"You can't believe that miracles happen. And I'm not talking time travel miracles: no one will believe that. I mean miracles of people healing from trauma. Like the drug addicts and racists and God and gun addicts you're angry at. You can't imagine them healing. If you could, you wouldn't be shaming them and writing them off like you are. You would have quite different ideas about how to prevent climate change."

"Oh please! Miracles of personal change?" I was on familiar ground now. "You came to tell me to listen to my *mother*? I was raised on endless miracle stories of individual change. The kind they have in Alcoholics Anonymous, but ours were geared towards ending conflicts. Do you know about my background?"

"I know enough to know you haven't learned the right lessons from it about social change."

"Then let me explain. My mother's religious people did conflict resolution behind the scenes in major conflicts. They may actually have prevented a war or two. But they didn't change poverty, they didn't change capitalism. They claimed to be "the Answer" for the world, but they had no answer to any of the powerful who refused to be one of their miracles.

Force is the only thing that moves the powerful. Violent force or nonviolent force: one or the other. Movements of the people to bring down the mighty! How do you think we got such democracy as we have?" I glared at her. She faced me back with Deliah's challenging eyes.

"Thank you for that demonstration of why Deliah left you: it wasn't mainly because of your white male entitlement, as painful as that was to her. She loved you in spite of it. And she did forgive you for the one time you were violent with her. She knew you wouldn't do it again."

Tears came to my eyes suddenly. "How do you know that?"

"She knew there was a lot of good in you. What she couldn't stand was the way you were always blaming and shaming your opponents. You leftwing men: always pointing the finger, never taking accountability."

"That's crazy! How do you know that about Deliah? Where is she?"

"Now, in 2022? I don't know. The last trace we have of her was in 1984, in Santa Fe. Her journals stopped then. Anyway, I came to talk politics with you. I've read online essays you wrote, the ones that went viral. You have a bigger influence than you realize, or you will do. After you died, even though you had changed your tune, people unearthed your old essays. The worse things got, the more people turned to those old writings, blaming the believers for almost everything, them and the fat cats, you called them, the billionaires, the capitalists. 'God and money, the two worst things to worship,' you wrote."

"Were you spying on me? I haven't published that phrase yet. That's in handwritten scribbles I jotted down this week."

"Really? It was in the notes the team gave me yesterday. They didn't imagine I could sway you, you know."

"Sway me? Young lady, I'm one of the strongest voices we've got for rolling back climate change! I'm attacking these idiots who deny it every day! Racists! Homophobes! The folk who suck up to the billionaires when they should be fighting them! Don't tell me you're soft on them?"

"That's the problem, Uncle Morris, right there. Can I call you uncle?"

"*What*'s the problem? And no, you can't."

"Oh, sorry, I'll try to remember. We call our elders Aunty and Uncle. I'm talking about how when change happens too fast and the old ways get thrown into the fire, people have to be *loved* to be brought along. You have to fight and organize to resist their politics, so in that sense force is part of it, but if they feel you don't love and respect them they'll just dig in, or retaliate."

"How can you love someone who believes white people are superior? Who thinks God made it so? Who thinks evolution never happened? But who takes meds based on modern biochemistry, for which evolution is central? They fly in airplanes but don't believe in science! They don't... Don't get me started!"

"But let me ask you this, Uncle Morris. How can I love someone who still uses a flush toilet in the California drought, who eats animals, who eats produce from soil-killing chemical agribusiness, who still drives a gas-powered car, and who speaks hate at people who are much more like him than they are like me? You should know better!

"And aside from that, don't you sometimes feel like you are hating your younger self, the self that at eighteen years old defended the Vietnam War in his school magazine?"

"My God, you've done your research, young lady."

"Well, I'm a historian. It's what we do."

"I'm not proud of that magazine piece, I can tell you."

"But Uncle Morris, don't be ashamed. You were eighteen. Raised conservative. You've learned. We're all learning."

"So who told you about that article?"

"You wrote about it in your journals, earlier this year, I believe. I read them in the vault."

"Vault? I've never let anyone see those journals and I don't have a vault." It began to strike me that something very, very odd was happening. Was she actually from the future?

"You didn't tell me your name."

"I'm Atsa."

"But that's my grandson's name."

"Of course. And I want to meet him. He's the other reason I came."

"Then let's go meet him." Now I was inviting her to my home?

We walked out onto Market Street towards the underground BART station. I found myself needing to prepare her for the encounter.

"My son and his wife, Atsa's parents, both died of Covid, you know that? Both Atsa's grandmothers previously dead from cancer, his other grandfather, Dr. John Begay, a thoracic surgeon in Albuquerque, completely absorbed in the pandemic. God knows why Atsa came to me after his parents died. He was a hell of a mess, still is. In and out of juvie. Fighting, drugs. I believe Begay was a kind grandfather, but overwhelmed after his wife died and the pandemic hit. Begay told Atsa he had to get to know the other side of his family. So he came. Berkeley's repute as a counterculture scene surely helped."

It was strangely natural to tell this woman stuff I hadn't shared with anyone else. "So it's up to me. And what do I know about raising a teenage boy? I wasn't even present for my own boy. Let alone a closed-up, furiously angry one like Atsa. Taking God knows what drugs. I've paid for therapy for him. He hates the therapist. It's a godawful mess. To be honest, he's a millstone round my neck just when I'm making maximum impact online and on the talkshows. I know I shouldn't talk about him like that. But he's not even trying."

She looked sympathetic as we walked but she was also short of breath, almost as if she was in a panic. As I watched, she took deep breaths and calmed herself again. A strange woman.

We took the escalator down to the BART station. "Do they still have these in your time?" I asked, meaning the underground trains. I could hear the ironic tone in my voice, as if I was only playing her game of visiting from the future. How hard it is to come out as a believer.

"Underground trains yes, the few that weren't flooded out. But these poor souls? No." She pointed to the bodies in blankets at the side of the tunnel, and the man trying to play a flute, too out of his mind to make the tune. "In our time, we look after each other. Ours is much more of a caring society."

On the train I quizzed her about what she meant by a caring society, and about how much tech *had* survived. She described how labor-intensive agriculture had become the norm, saving fuel and minerals, ensuring healthy lives, enabling community. Travel was way down. People-centered and -owned media networked the world. I wondered what the passengers around us thought we were doing: a kind of improv, acting roles to come up with a screenplay?

I live up in the twisting streets that follow the contours of the Berkeley hills. It was a good walk from Downtown Berkeley BART. Normally I'd take a cab, never an Uber, given their labor practices. But after her work on it, my back was feeling amazingly good. "Let's walk," I said, bravely. She grinned. We needed to take two of the long staircases between streets and I was utterly out of breath and sweating by the time we reached my home. She did it easily, her anxiety once more in abeyance. So odd.

The house is drab, for lack of paint and any architectural input whatsoever. Still, it's worth millions for its view of the Bay and the Golden Gate. "I inherited it from my parents, who bought it cheap," I told her, fearing she'd think me one of the fat cats I excoriated. Which of course I am, compared to many folk at the bottom of the hill, especially those whose skin color was used to keep their parents and grandparents out of the housing market. I unlocked the front door, wondering whether the next generation would be home, the boy I was leaving the house to. We were met by silence, no pounding hiphop.

I turned on the electric kettle. "One hundred percent renewable power," I felt obliged to say in my defense. "I pay extra for it."

Just then the front door slammed. "Guess who," I said.

A grim teenage boy strode in. He had his mother's brown skin, high cheekbones, and sturdy build. His tightly muscled look was his own creation, as were the black half-moons under his eyes. A sleepless gym rat.

"Atsa, meet Atsa," I said.

"Really? I never met another Atsa up this way. Who are you?" That was a lot of words for my Atsa. He was staring at her like she had stepped right out of one of his comic books: I'd never seen him so interested in a live human.

"*Yá'át'ééh*," she said.

"*Yá'át'ééh*."

The only Navajo word I knew. And then they were talking vigorously in his native language.

This was a different boy than the one I knew. Eager to make contact with her.

"Please guys, speak English. I want to know what's going on."

He stared at me. Reduced to his customary silence. I regretted my interruption.

Then he turned to her and asked, "Are you staying?"

"Do you want me to?"

He considered it carefully.

She added. "I want to introduce you to someone I think you'll appreciate."

"Ok."

What was going on here? It was like they were mind-melding. The strangest day of my life was getting stranger. Maybe she could work a miracle with Atsa. Help him feel life was worth living. I certainly couldn't.

"I've got a headache," I complained. "I'm going to rest my back." The pain had started up again. The only surprise was how long her woo-woo healing had lasted. They'd get on better without me, anyway.

Asleep on the couch, I dreamed of addicts injecting each other in a back alley behind a restaurant, packed with stinking garbage. They were coughing their lives out with Covid. Their baby boy cried helplessly in a dirty diaper, the mother trying so hard to help him while she died. I woke sweating and unable to get the horror from my mind. I struggled up despite the pain in my back, tried to swallow a Vicodin dry, staggered to the kitchen faucet, drank from a dirty mug, and put on the kettle. The dream stayed with me, as vivid as the reality it reflected. It was all true but for the baby. Atsa was already fourteen and living with his Begay grandfather, when his parents died in that alley. My son, Christopher. His longtime love, Angelina. I should pray for forgiveness. For all the ways I must have failed Chris. Even though I still don't really understand what I had done. My divorce, of course. But lots of kids' parents divorce: it doesn't turn them into addicts. Many times in the past year I had longed for the God of my childhood, the God I had believed in, who could forgive, and understand, and guide. But I had never felt love from that God. Only demands for obedience.

Like I had never really felt love from my mother. I had known she loved me. But that was head knowledge. She had been too repressed, too upper class, too faded WASP gentry, to hug me. And despite my rebellion, my immersion in the 70s counterculture, my certainty that my wife, Rose, and I would do a better job, we had somehow failed our own son. Over the past year I had frequently thanked the cosmos that cancer had taken Rose before Chris and Angelina died.

Hours later, I was back at my laptop excoriating "environmentalists" who imagined the corporate world could ever be a driving force for climate sanity, when I heard chattering voices coming up the steps to the front door. My Atsa, chattering? The other Atsa, talking quietly. I needed a name to distinguish her from the boy: Miracle Atsa? And then a third voice! A voice I hadn't heard in over thirty years. It couldn't be. Not possible. Those Boston Brahmin vowels.

The two Atsas entered, still talking, and then her. I stood up.

"Oh darling!" she exclaimed. "You look so old! Older than me!"

"Who are you? I mean, I m… m… mean…"

"Oh darling, you haven't stuttered since you were eight years old!"

We stared at each other. She looked in her fifties, as she had been during my late teens. She wore her stolid brown brogues, nylons, conservative tweed skirt, cashmere twin set, pearl necklace, pearl earrings, permed hair.

I turned to the time traveler: "What have you done?"

"You two need to talk."

"You can do this? Am I hallucinating?"

"No, you're not, darling," Mother said. "I'm really here. I think. Atsa says it's time travel, not a hallucination. But does it matter? Atsa's right. We need to talk."

"What good did talking ever do? You never listened."

"I listened. But I didn't understand. Maybe I can now."

I held my head and looked at the floor. Took three long deep breaths to calm down. Studied the cracks between the old floorboards. I was clearly overwrought. In need of a new psych with better meds. I looked up. She was still there.

"Why the hell would you understand now? Look at you." The perm, the pearls. "You're the age when all you did was wail that I was abandoning God."

"And look at you! I can see what you've made of yourself, and the Atsas have told me more. I'm sorry. I'm truly sorry."

"What for?"

"I failed you. I know that, or you wouldn't be spewing hate towards the haters. When I met the Oxford Group before the War, I was a far right racist. All my family were, as you know. My father opposed FDR fiercely every day of his tenure in the House. Of course, I never voted for FDR. We knew all about his philandering. But I did vote for Truman. Think about that. Do you know why?"

"Because Truman supported the Group."

"No, it was because I had changed. I'll use Atsa's word: I had healed. I had faced my own bitterness towards the clergyman who raped me."

"What? I never knew about that."

"No of course you didn't. It wasn't anything I could speak about. Atsa has told me I need to do so now, for you to understand me. You see, I felt sure that the rape was my fault. I was a sinful girl, full of lust and ambition. And afterwards, I was consumed by anger and self-loathing in equal parts. I started drinking too much. But then I went to that Group houseparty, and Jesus came into my life. I gave my life to Him and I became free. I became honest about my sins, I gave my life over to God. And then my sins didn't have the power over me they had before. I became part of a band of friends and warriors, out to heal the world. And then my politics changed. I was open to whoever had the most integrity."

"You voted for Eisenhower. And Nixon."

"I know, voting for Nixon was a mistake. But Nixon was a family man while that Jack Kennedy was a worse philanderer than FDR."

"What is this, Miracle Atsa? Or shall I call you Nightmare Atsa? You're trying to work some bizarre political reeducation in me by having my mother tell me about her sins and her voting record? Mother, what you people in MRA never understood is that it's systems of oppression that matter more than the marital fidelity of politicians!"

"Hold on. What's MRA?" young Atsa asked. I looked at him for the first time since Mother had walked into the room. His eyes, which had been so dead all this past year, were alive with curiosity. No wonder: he was getting the family dirt. "Great Grandma, that was an awesome story about how you recovered from being raped. That's trauma recovery, big time. My friend June was raped and she needs something like that. So Grandad, you never told me about this MRA. What was it?"

"God! I hated that question when I was a teen. I couldn't explain what MRA was and no one understood when I tried." I sat down abruptly. "Mother, sit, all of you, sit." I should have been the host, got them glasses of water at least. But I was too bemused. I looked at young Atsa again, caught by the curiosity in his face. "OK. MRA was short for Moral Re-Armament. Let's say it was a cross between AA, Scientology, and a conflict mediation service. And throw in hefty doses of evangelical revivalism and pan-religious kumbaya."

"Sounds weird."

"Morris, you can do better than that," my mother said in her severest tone, as if I was ten years old.

I stared at her and decided young Atsa did deserve better, and maybe Mother did too. "Ok, I'll try. This is part of your heritage, Atsa. It's what fucked me up. But I'll do my best to be fair. Interrupt me if I go on too long. In the 1930s there was this thing called the Oxford Group. It was a fundamentalist Christian revival focused on the experience of personal change. Its greatest result was to spawn Alcoholics Anonymous, which is all about personal change. But the Group had absurd ambitions. It thought it could stop Hitler by a revival of personal change. In 1938 it renamed itself Moral Re-Armament, or MRA: moral and spiritual rearmament instead of, or in addition to, military rearmament. After the war it was saddled with that awful name. Still, it built an unusual record of conflict resolution in high places. Its post-war reconciliation work between French and Germans was supposedly critical to founding the institution that became the European Union."

Young Atsa's eyes were starting to glaze over. How could I distill a lifetime's experience for my grandson? "Here's the thing. No one knows about MRA today. It's been totally ignored and forgotten by academia and the media because it was homophobic, cultic, anti-sex, anti-intellectual, anti-Left: it thought personal change was all that was needed to end hunger, war, and conflict. My teen friends hated all the stuff about sin and obedience to God and redemption. How am I doing, Mother? Is that about right?" All the old fury was resurfacing.

"You're bitter. Why? I don't understand. You grew up in a loving home."

"A loving home being one in which no one ever hugged anyone? I can't even recall you touching me, Mother."

"Oh! Oh, my dear. That was just our way. That was New England Brahmin culture. That was..."

"Breathe, Morris. Breathe," said Miracle/Nightmare Atsa. And then, as if to give me time to do so, she asked Mother, "Shall I tell you what I think your not hugging him was?"

"Yes, dear, please enlighten me."

"I think it was white trauma."

"Hey, tell me about that," young Atsa asked.

"Christians are all about sin as the problem, right? But what if the problem is really trauma?"

"You mean we do bad shit not because we're bad, but because we've been hurt."

"Exactly. When we're hurt traumatically, it changes us. We can act defensively, or lash out, or numb ourselves: everyone's different. You're lucky to be living in the era when trauma healing is starting to become known. In your time, today, restorative justice is gaining traction. That's based on the idea that people who do bad stuff need healing not shaming and punishment. That's what you needed, Atsa, when you were acting out: healing not juvie."

"On the Res they call it Indigenous Peacemaking. My family hasn't had much to do with it. I mean, I've heard about intergenerational trauma: everyone on the Res is suffering from that. That's what my grandparents escaped from. Much good it did my mom."

"In your time people are just starting to understand how intergenerational trauma can affect the descendants of slavery and genocide. But if it's hurt people who hurt people, then why were the white people committing those crimes against humanity? Had they been hurt? Yes. That's what I'm saying. White trauma. Only they were in charge, so no one punished them. No one tried to heal them either, unless it was people like Morris's mother."

"Thank you, my dear," Mother said. "Thank you. We were trying. From what you've been telling me, I do understand how little we understood about some people, like the homosexuals."

"In MRA you did a lot of confessing your sins, right? Well restorative justice involves accountability, admitting what you did, which is a lot like confessing your sins. But then, with a trauma perspective, accountability requires working out why you did it, how you had been hurt yourself, probably in childhood, and how you had been taught a toxic culture. And from there, how to heal yourself."

"So you're trying to tell me," I said to Miracle Atsa, "that trauma healing and restorative justice are like a secular replacement for MRA and AA?"

"In our time, we call it Healing Community, and it's not just secular. It's spiritual too."

Young Atsa asked, "How were white people hurt? You're just making excuses for them."

"Well, some historians in your time began asking why had white people colonized the world? Was it just because they could, because they had the weaponry, had industrialized first? But why the brutality, why the contempt? And why did they have the weaponry, when others didn't? Well, they developed it by fighting each other. They'd been killing each other for centuries in Europe, before they ever ventured out. A third to a half of the Irish population died from Cromwell's repression, a third to a half of the German population died in the Thirty Years War. My own theory about why they fought each other harder and longer than people in other parts of the world is about their continent's geography, but that's for another day.

"Of course, there was so much more to European civilization than vicious warfare. All the art, music, theatre, spirituality, science. But there is a psychological cost to decimating the Cathars, the Irish, the Scots Highlanders, the Jews, the Slavs (that's where the word "slave" comes from), the enslaved Africans, the indigenous Americans and Australians, a cost to impoverishing India and China, the wealthiest parts of the world. So we can add to the traumas of Europe's self-harms the secondary traumas of killing and exploiting others. The emotional repression of the British, especially of the middle and upper middle classes which served the Empire, and their WASP descendants is one of those costs. That's what I meant."

"How can the trauma of being a colonial officer compare with being enslaved?"

"Well, of course it doesn't. But the white culture was all about training its sons to repress their emotions and tenderness so they could be brutal in the most civilized way. What white people need to do to stop their own racism is to heal themselves from the long history of traumas that built their abusive culture. That's all I meant. And it's not about shaming themselves and each other for being racist: that doesn't heal us. You know I'm partly white, like you, Atsa."

"My dad was such a mess," young Atsa said.

"So Trump voters are traumatized? *That's* why you're here?" I burst out. "To convert me to healing those racist, punitive morons on the Right? Because it's all about Post Traumatic Whiteness Syndrome? What a crock!"

"No darling, she's right. She wants you to become a healer, not a divider. And I'm starting to see what went wrong in our family. I am so sorry."

"What for? Being part of that religious cult?"

"No, dear. And you know quite well it was not a cult. Well, not much of one. What I'm sorry for is… I have always felt there is something horrid about our type of people, how we can't express love like other people. Like our Mexican maid, for example, the way she adored her kids. We're repressed, I do know that. I didn't feel able to change it. And then, when you were a baby the parenting experts said that too much mothering and tenderness made our boys weak, or homosexual."

"And what would have been wrong with that? Oh, forget it, we can't relitigate my whole experience of being raised by WASPs in the 50s and 60s."

"I never was entirely at ease with it," Mother said. "But you have to understand the times, and… No. If I have learned one thing about making an apology it's not to put in the self-defense. Morris, will you please stand up?"

Bemused, I did as she asked. I didn't know why.

Then she stepped resolutely forward, opened her arms and drew me into a very gentle and loving hug. The hug I had longed for all through my childhood.

I froze.

"It's all right, darling. Believe me, I've wanted to do this for ages." To my own shock, my head dropped onto her shoulder and I began to weep. My chest heaved with sobs.

"I'm so sorry," she said. "I'm so sorry. There, there, darling. Just cry."

When I finally lifted my head off my mother's tear-sodden shoulder, I felt I could speak. But I had nothing to say. Would I accuse her of not doing that before? No, I just felt grateful and bewildered. And, somehow, unmanned.

So I turned away from her and said, "Miracle Atsa—that's what I'm calling you in my mind—what the fuck are you doing? This is enough to drive a man insane." I sat back down, hard.

"Grandpa," young Atsa broke in, "that was awesome."

"What was?"

"Seeing you cry. Never thought I'd see it." His face was younger and more eager than it had been all this past long year since he came to me. That stopped my throat. I looked accusingly at Miracle Atsa, and wagged my finger at her. Her fault. I felt like I could weep and never stop. I held it back and croaked, "Why? Why is it important to see me cry?"

"Grandpa, no offense, but you're an old *biligáana* full of rage. My other grandfather says you need healing. I think it's begun."

"Goddammit, is the whole family on my case? And what the hell's a bill o' garna?"

"A white person, Grandad, just a white person."

Miracle Atsa smiled as she wagged a finger at me. "Loretta Ross, who we call the grandmother of the new culture that was just beginning to arise among activists in your time, used to say that the way white leftists put each other down for not being woke enough was just another form of white supremacy: blaming and shaming is central to white supremacy. And you're a master at it, Morris Dale."

"Whoa. Awesome." said young Atsa.

"Darling boy," my mother said, and she was not referring to the teenager in the room. "Let's have a cup of coffee. Or tea. What do you prefer these days?"

The others left me alone while young Atsa showed his great grandmother how to make coffee and Miracle Atsa went out to the back yard: to breathe, she said. Sitting on the couch I stewed in a morass of confusion. Young Atsa's face kept coming back to me. I had never really given much thought to his Navajo grandparents: they had seemed so Westernized to me: a surgeon, after all, and a data statistician. What did they really think of white people? Chris had kept Angelina away from me, so I had only the slightest idea about her.

Mother and Atsa took a tray of coffees out to the backyard. I followed. It was a wild place, untamed since Mother's death. I called it the wilderness. It was a home now for deer that leapt the fences, fawned in the back corner, and ate the plums in season. I brought out a deck chair, slumped in it, then got up to lay on my back in the long grass: aching again. I sank upwards into the deep blue of the sky.

I fell asleep there, and woke to find Miracle Atsa kneeling over me again, hands palm down over my stomach, no doubt "moving my energies" or some other nonsense. Except that my pain had gone again.

"Tell me what's going to happen," young Atsa said. "You come from the future, which means there really is a future? Everything I'm listening to says it's all going to be dark and brutal."

Miracle Atsa took a huge breath and said, "It's up to you." The words came out of her with such gravity, such certainty, it seemed to me that she literally meant it was up to young Atsa. "I can teach you," she said. "You can probably find Loretta Ross and go study with her. And go find the Indigenous Peacemakers. And you could do worse than read *The Ministry of the Future*, by Kim Stanley Robinson. He didn't get it all right, how could he; and he didn't include Loretta, or restorative justice, which he could have done; but he was closer than most futurists in your time. How a culture swings from hatred to cooperation: it depends on the visionaries, the people who see the dangers, but it takes more than prophesies of doom, and more than dreaming of spiritual revolution. The visionaries have to work out the actual technology, the practices, the how tos, the speech patterns, for calling people in, not calling them out. New cultures only ever start with small numbers who are living in a new way."

"That's what I want," said young Atsa. "That's what I want."

2142

"Oh, thanks be!" Maria dropped their hoe and stood unmoving in shock. "When you left for the Physics Lab yesterday I thought…" They began to cry.

"Hush, I'm here," Atsa took Maria in their arms. "I'm here."

"You didn't go after all? They told me you had."

"I went. I spent weeks with Atsa Begay and Morris and his mother, Mary. Then I came back to today."

"Maiden, Mother and Crone. Let me look at you" Maria pushed Atsa away. "Your face is different."

"It was amazing, Maria. I was so anxious. But I calmed down once we were all talking and healing together."

"And you claim not to be a healer."

"It turns out that I have learned a few things."

"So tell me. Was it important like your grandmother told you it would be? Is it because of you that Atsa Begay became the leader and speaker and singer that they were? Quoting your own book, 'Begay was perhaps the most effective leader of their time at calling in Americans, right across the political spectrum, to creating the kinship society?'"

"I believe so, I truly believe so. Am I forgiven?"

"Come here, sweetheart. I love you." Maria opened their arms. "Just don't ever do it again."

Index[1]

[1] Note: Page numbers followed by 'n' refer to notes.

J. L. Hennessey (ed.), *History and Speculative Fiction*,
https://doi.org/10.1007/978-3-031-42235-5